GACE Physics
030
031
Teacher Certification Exam

By: Sharon Wynne, M.S.
Southern Connecticut State University

"And, while there's no reason yet to panic, I think it's only prudent that we make preparations to panic."

XAMonline, INC.
Boston

Copyright © 2007 XAMonline, Inc.
All rights reserved. No part of the material protected by this copyright notice may be reproduced or utilized in any form or by any means, electronic or mechanical, including photocopying, recording or by any information storage and retrievable system, without written permission from the copyright holder.

To obtain permission(s) to use the material from this work for any purpose including workshops or seminars, please submit a written request to:

XAMonline, Inc.
21 Orient Ave.
Melrose, MA 02176
Toll Free 1-800-509-4128
Email: info@xamonline.com
Web www.xamonline.com
Fax: 1-781-662-9268

Library of Congress Cataloging-in-Publication Data

Wynne, Sharon A.
    Physics 030, 031: Teacher Certification / Sharon A. Wynne. -2nd ed.
    ISBN 978-1-58197-569-7
    1. Physics 030, 031.   2. Study Guides.   3. GACE
    4. Teachers' Certification & Licensure.   5. Careers

Disclaimer:
The opinions expressed in this publication are the sole works of XAMonline and were created independently from the National Education Association, Educational Testing Service, or any State Department of Education, National Evaluation Systems or other testing affiliates.

Between the time of publication and printing, state specific standards as well as testing formats and website information may change that is not included in part or in whole within this product. Sample test questions are developed by XAMonline and reflect similar content as on real tests; however, they are not former tests. XAMonline assembles content that aligns with state standards but makes no claims nor guarantees teacher candidates a passing score. Numerical scores are determined by testing companies such as NES or ETS and then are compared with individual state standards. A passing score varies from state to state.

Printed in the United States of America

GACE: Physics 030, 031
ISBN: 978-1-58197-569-7

TEACHER CERTIFICATION STUDY GUIDE

Table of Contents

SUBAREA I.   MECHANICS

COMPETENCY 1.0   UNDERSTAND MOTION IN ONE AND TWO DIMENSIONS 1

Skill 1.1   Calculating displacement, average velocity, instantaneous velocity, and acceleration in a given frame of reference 1

Skill 1.2   Solving problems involving displacement, time, velocity, and constant acceleration 2

Skill 1.3   Interpreting algebraically and graphically relationships among position, velocity, acceleration, and time 4

Skill 1.4   Analyzing problems involving motion in two dimensions 5

Skill 1.5   Analyzing properties of vectors and solving problems involving vector quantities analytically and graphically 8

COMPETENCY 2.0   UNDERSTAND NEWTON'S LAWS AND THE UNIVERSAL LAW OF GRAVITATION 12

Skill 2.1   Analyzing examples of Newton's laws of motion in daily life 12

Skill 2.2   Applying knowledge of free-body diagrams and vector properties to solve problems 14

Skill 2.3   Analyzing and solving problems involving frictional forces and coefficients 15

Skill 2.4   Solving problems involving gravitational forces 16

Skill 2.5   Applying knowledge of Newton's laws of motion to solve problems in one and two dimensions 17

Skill 2.6   Solving problems involving torque and static equilibrium 19

TEACHER CERTIFICATION STUDY GUIDE

COMPETENCY 3.0 UNDERSTAND THE CONSERVATION OF ENERGY 23

Skill 3.1 Calculating the kinetic and potential energy of mechanical systems 23

Skill 3.2 Applying knowledge of the law of conservation of energy and the work-energy theorem to solve problems involving conservative and nonconservative forces 24

Skill 3.3 Analyzing mechanical systems in terms of work, power, and energy 28

COMPETENCY 4.0 UNDERSTAND MOMENTUM AND ITS CONSERVATION 30

Skill 4.1 Applying knowledge of the concept of impulse and the conservation of momentum to solve problems in one and two dimensions 30

Skill 4.2 Applying knowledge of the concepts of energy and momentum to analyze elastic and inelastic collisions 31

Skill 4.3 Applying knowledge of vectors to solve momentum problems 32

Skill 4.4 Analyzing examples of conservation of angular momentum in everyday life 33

TEACHER CERTIFICATION STUDY GUIDE

SUBAREA II. WAVES AND THERMAL ENERGY

COMPETENCY 5.0 UNDERSTAND THE CHARACTERISTICS OF MECHANICAL WAVES 36

Skill 5.1 Analyzing models of harmonic motion 36

Skill 5.2 Analyzing the production and propagation of sound waves 38

Skill 5.3 Analyzing reflection and transmission of mechanical waves 40

Skill 5.4 Applying knowledge of the superposition principle to solve problems involving constructive and destructive interference 43

Skill 5.5 Analyzing waves and solving problems involving amplitude, wavelength, period, frequency, and propagation speed in various media 44

COMPETENCY 6.0 UNDERSTAND THE FUNDAMENTAL PRINCIPLES OF LIGHT AND OPTICS 47

Skill 6.1 Analyzing properties of images produced by various mirrors 46

Skill 6.2 Applying knowledge of ray diagrams and Snell's law to solve problems involving refraction 49

Skill 6.3 Analyzing properties or images produced by convex and concave lenses 52

Skill 6.4 Analyzing the phenomena of dispersion, diffraction, and polarization 54

Skill 6.5 Analyzing wave properties of the electromagnetic spectrum 56

PHYSICS

COMPETENCY 7.0	UNDERSTAND THE PRINCIPLES OF THERMODYNAMICS	59
Skill 7.1	Differentiating between thermal energy and temperature and solving problems involving thermal energy	59
Skill 7.2	Identifying methods of thermal energy transfer	59
Skill 7.3	Applying knowledge of thermodynamic work and the law of conservation of energy to solve a variety of problems	60
Skill 7.4	Demonstrating knowledge of the second law of thermodynamics	61
Skill 7.5	Analyzing the states of matter and energy transformation during phase changes	63

TEACHER CERTIFICATION STUDY GUIDE

SUBAREA III. ELECTRICITY, MAGNETISM, AND ATOMIC PHYSICS

COMPETENCY 8.0 UNDERSTAND ELECTRIC CHARGE AND INTERACTIONS BETWEEN CHARGED OBJECTS 66

Skill 8.1 Demonstrating knowledge of principles and application of electrostatics 66

Skill 8.2 Solving problems using Coulomb's law 68

Skill 8.3 Demonstrating knowledge of the electric field in the vicinity of point charges 70

Skill 8.4 Solving basic problems involving electrostatic potential and electrostatic potential energy 71

COMPETENCY 9.0 UNDERSTAND MAGNETS, MAGNETIC FIELDS, AND ELECTROMAGNETIC INDUCTION 73

Skill 9.1 Demonstrating knowledge of the properties of permanent magnets . 73

Skill 9.2 Determining the strength and orientation of the magnetic field near a current-carrying wire 74

Skill 9.3 Solving problems involving charged particles moving through a magnetic field 75

Skill 9.4 Demonstrating knowledge of direction and relative magnitude of an induced EMF in a conductor 76

Skill 9.5 Analyzing application of electromagnetism 79

Skill 9.6 Demonstrating knowledge of the generation of electromagnetic waves and their applications 82

COMPETENCY 10.0	UNDERSTAND THE PROPERTIES OF ELECTIC CIRCUITS 84
Skill 10.1	Interpreting simple schematic diagrams of DC circuits 84
Skill 10.2	Analyzing series and parallel circuits using Ohm's law 85
Skill 10.3	Demonstrating knowledge of energy conservation in simple circuits 85
Skill 10.4	Differentiating between alternating and direct current circuits 87

COMPETENCY 11.0	UNDERSTAND THE BASIC PROCESSES OF ATOMIC AND NUCLEAR PHYSICS 88
Skill 11.1	Analyzing differences between fission and fusion and the applications of each 88
Skill 11.2	Demonstrating knowledge of models of nuclear and subatomic structures and behaviors 91
Skill 11.3	Demonstrating knowledge of the half-life of radioactive isotopes 93
Skill 11.4	Demonstrating knowledge of how the basic principles of quantum mechanics can be used to describe the properties of light and matter 94

SUBAREA IV.   **CHARACTERISTICS OF SCIENCE**

COMPETENCY 12.0   **UNDERSTAND THE CHARACTERISTICS OF SCIENTIFIC KNOWLEDGE AND THE PROCESS OF SCIENTIFIC INQUIRY** **101**

Skill 12.1   Demonstrating knowledge of the nature, purposes, and characteristics of science 101

Skill 12.2   Recognizing the dynamic nature of scientific knowledge through continual testing, revision, and the occasional rejection of existing theories 102

Skill 12.3   Determining an appropriate scientific hypothesis or investigative design for addressing a given problem 103

Skill 12.4   Demonstrating knowledge of the principles and procedures for designing and carrying out scientific investigations 106

Skill 12.5   Recognizing the importance of and strategies for avoiding bias in scientific investigations 107

TEACHER CERTIFICATION STUDY GUIDE

COMPETENCY 13.0   UNDERSTAND THE COLLECTION, ANALYSIS, AND COMMUNICATION OF SCIENTIFIC DATA 109

Skill 13.1   Identifying appropriate tools and units for measuring objects of substances 109

Skill 13.2   Recognizing potential safety hazards and procedures for the safe and proper use of scientific tools, instruments, chemical, and other materials in investigations 112

Skill 13.3   Recognizing the concepts of precision, accuracy, and error and identifying potential sources of error in gathering and recording data 114

Skill 13.4   Identifying methods and criteria for organizing and analyzing data 116

Skill 13.5   Identifying appropriate methods for communicating the outcomes of scientific investigations 118

Skill 13.6   Demonstrating familiarity with effective resources and strategies for reading to gain information about science-related topics and developing subject-area vocabulary 121

Sample Test 123

Answer Key 145

Rigor Table 146

Rationales with Sample Questions 147

TEACHER CERTIFICATION STUDY GUIDE

Great Study and Testing Tips!

What to study in order to prepare for the subject assessments is the focus of this study guide but equally important is *how* you study.

You can increase your chances of truly mastering the information by taking some simple, but effective steps.

Study Tips:

1. Some foods aid the learning process. Foods such as milk, nuts, seeds, rice, and oats help your study efforts by releasing natural memory enhancers called CCKs (*cholecystokinin*) composed of *tryptopha*n, *choline*, and *phenylalanine*. All of these chemicals enhance the neurotransmitters associated with memory. Before studying, try a light, protein-rich meal of eggs, turkey, and fish. All of these foods release the memory enhancing chemicals. The better the connections, the more you comprehend.

Likewise, before you take a test, stick to a light snack of energy boosting and relaxing foods. A glass of milk, a piece of fruit, or some peanuts all release various memory-boosting chemicals and help you to relax and focus on the subject at hand.

2. Learn to take great notes. A by-product of our modern culture is that we have grown accustomed to getting our information in short doses (i.e. TV news sound bites or USA Today style newspaper articles.)

Consequently, we've subconsciously trained ourselves to assimilate information better in neat little packages. If your notes are scrawled all over the paper, it fragments the flow of the information. Strive for clarity. Newspapers use a standard format to achieve clarity. Your notes can be much clearer through use of proper formatting. A very effective format is called the *"Cornell Method."*

> Take a sheet of loose-leaf lined notebook paper and draw a line all the way down the paper about 1-2" from the left-hand edge.
>
> Draw another line across the width of the paper about 1-2" up from the bottom. Repeat this process on the reverse side of the page.

Look at the highly effective result. You have ample room for notes, a left hand margin for special emphasis items or inserting supplementary data from the textbook, a large area at the bottom for a brief summary, and a little rectangular space for just about anything you want.

3. **Get the concept then the details.** Too often we focus on the details and don't gather an understanding of the concept. However, if you simply memorize only dates, places, or names, you may well miss the whole point of the subject.

A key way to understand things is to put them in your own words. If you are working from a textbook, automatically summarize each paragraph in your mind. If you are outlining text, don't simply copy the author's words.

Rephrase them in your own words. You remember your own thoughts and words much better than someone else's, and subconsciously tend to associate the important details to the core concepts.

4. **Ask Why?** Pull apart written material paragraph by paragraph and don't forget the captions under the illustrations.

Example: If the heading is "Stream Erosion", flip it around to read "Why do streams erode?" Then answer the questions.

If you train your mind to think in a series of questions and answers, not only will you learn more, but it also helps to lessen the test anxiety because you are used to answering questions.

5. **Read for reinforcement and future needs.** Even if you only have 10 minutes, put your notes or a book in your hand. Your mind is similar to a computer; you have to input data in order to have it processed. *By reading, you are creating the neural connections for future retrieval.* The more times you read something, the more you reinforce the learning of ideas.

Even if you don't fully understand something on the first pass, *your mind stores much of the material for later recall.*

6. **Relax to learn so go into exile.** Our bodies respond to an inner clock called biorhythms. Burning the midnight oil works well for some people, but not everyone.

If possible, set aside a particular place to study that is free of distractions. Shut off the television, cell phone, and pager and exile your friends and family during your study period.

If you really are bothered by silence, try background music. Light classical music at a low volume has been shown to aid in concentration over other types. Music that evokes pleasant emotions without lyrics is highly suggested. Try just about anything by Mozart. It relaxes you.

7. Use arrows not highlighters. At best, it's difficult to read a page full of yellow, pink, blue, and green streaks. Try staring at a neon sign for a while and you'll soon see that the horde of colors obscure the message.

A quick note, a brief dash of color, an underline, and an arrow pointing to a particular passage is much clearer than a horde of highlighted words.

8. Budget your study time. Although you shouldn't ignore any of the material, *allocate your available study time in the same ratio that topics may appear on the test.*

TEACHER CERTIFICATION STUDY GUIDE

Testing Tips:

1. Get smart, play dumb. Don't read anything into the question. Don't make an assumption that the test writer is looking for something else than what is asked. Stick to the question as written and don't read extra things into it.

2. Read the question and all the choices *twice* before answering the question. You may miss something by not carefully reading, and then re-reading both the question and the answers.

If you really don't have a clue as to the right answer, leave it blank on the first time through. Go on to the other questions, as they may provide a clue as to how to answer the skipped questions.

If later on, you still can't answer the skipped ones . . . *Guess.* The only penalty for guessing is that you *might* get it wrong. Only one thing is certain; if you don't put anything down, you will get it wrong!

3. Turn the question into a statement. Look at the way the questions are worded. The syntax of the question usually provides a clue. Does it seem more familiar as a statement rather than as a question? Does it sound strange?

By turning a question into a statement, you may be able to spot if an answer sounds right, and it may also trigger memories of material you have read.

4. Look for hidden clues. It's actually very difficult to compose multiple-foil (choice) questions without giving away part of the answer in the options presented. In most multiple-choice questions you can often readily eliminate one or two of the potential answers. This leaves you with only two real possibilities and automatically your odds go to Fifty-Fifty for very little work.

5. Trust your instincts. For every fact that you have read, you subconsciously retain something of that knowledge. On questions that you aren't really certain about, go with your basic instincts. **Your first impression on how to answer a question is usually correct.**

6. Mark your answers directly on the test booklet. Don't bother trying to fill in the optical scan sheet on the first pass through the test.

Just be very careful not to miss-mark your answers when you eventually transcribe them to the scan sheet.

7. Watch the clock! You have a set amount of time to answer the questions. Don't get bogged down trying to answer a single question at the expense of 10 questions you can more readily answer.

PHYSICS

TEACHER CERTIFICATION STUDY GUIDE

SUBAREA I.   MECHANICS

COMPETENCY 1.0   UNDERSTAND MOTION IN ONE AND TWO DIMENSIONS

Skill 1.1   Calculating displacement, average velocity, instantaneous velocity, and acceleration in a given frame of reference

Kinematics is the part of mechanics that seeks to understand the motion of objects, particularly the relationship between position, velocity, acceleration and time.

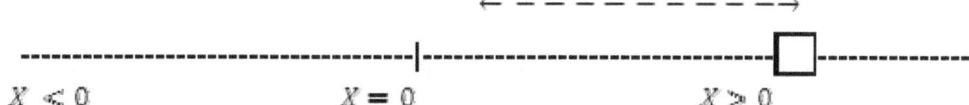

$x < 0$ \qquad $x = 0$ \qquad $x > 0$

The above figure represents an object and its displacement along one linear dimension.

First we will define the relevant terms:

1. Position or Distance is usually represented by the variable *x*. It is measured relative to some fixed point or datum called the origin in linear units, meters, for example.

2. Displacement is defined as the change in position or distance which an object has moved and is represented by the variables D, d or Δx. Displacement is a vector with a magnitude and a direction.

3. Velocity is a vector quantity usually denoted with a V or v and defined as the rate of change of position. Typically units are distance/time, m/s for example. Since velocity is a vector, if an object changes the direction in which it is moving it changes its velocity even if the speed (the scalar quantity that is the magnitude of the velocity vector) remains unchanged.

   i) Average velocity: $\vec{v} = \frac{\Delta d}{\Delta t} = (d_1 - d_0)/(t_1 - t_0)$.
   The ratio, $\Delta d/\Delta t$ is called the average velocity. Average here denotes that this quantity is defined over a period Δt.

   ii) Instantaneous velocity is the velocity of an object at a particular moment in time. Conceptually, this can be imagined as the extreme case when Δt is infinitely small.

4. Acceleration represented by *a* is defined as the rate of change of velocity and the units are m/s^2. Both an average and an instantaneous acceleration can be defined similarly to velocity.

PHYSICS

From these definitions we develop the kinematic equations. In the following, subscript *i* denotes initial and subscript *f* denotes final values for a time period. Acceleration is assumed to be constant with time.

$$v_f = v_i + at \qquad (1)$$

$$d = v_i t + \frac{1}{2} a t^2 \qquad (2)$$

$$v_f^2 = v_i^2 + 2ad \qquad (3)$$

$$d = \left(\frac{v_i + v_f}{2}\right) t \qquad (4)$$

In two dimensions the same relationships apply, but each dimension must treated separately.

Skill 1.2 Solving problems involving displacement, time, velocity, and constant acceleration

Simple problems involving distance, displacement, speed, velocity, and constant acceleration can be solved by applying the kinematics equations from the proceeding section. The following steps should be employed to simplify a problem and apply the proper equations:

1. Create a simple diagram of the physical situation.
2. Ascribe a variable to each piece of information given.
3. List the unknown information in variable form.
4. Write down the relationships between variables in equation form.
5. Substitute known values into the equations and use algebra to solve for the unknowns.
6. Check your answer to ensure that it is reasonable.

Example:
A man in a truck is stopped at a traffic light. When the light turns green, he accelerates at a constant rate of 10 m/s². **a)** How fast is he going when he has gone 100 m? **b)** How fast is he going after 4 seconds? **c)** How far does he travel in 20 seconds?

a=10 m/s²

Solution:
We first construct a diagram of the situation.
In this example, the diagram is very simple, only showing the truck accelerating at the given rate. Next we define variables for the known quantities (these are noted in the diagram):

a=10 m/s²; v_i=0 m/s

Now we will analyze each part of the problem, continuing with the process outlined above.

For part **a)**, we have one additional known variable: d=100 m

The unknowns are: v_f (the velocity after the truck has traveled 100m)

Equation (3) will allow us to solve for v_f, using the known variables:

$$v_f^2 = v_i^2 + 2ad$$

$$v_f^2 = (0m/s)^2 + 2(10m/s^2)(100m) = 2000\frac{m^2}{s^2}$$

$$v_f = 45\frac{m}{s}$$

We use this same process to solve part **b)**. We have one additional known variable: t=4 s

The unknowns are: v_f (the velocity after the truck has traveled for 4 seconds)

Thus, we can use equation (1) to solve for v_f:

$$v_f = v_i + at$$

$$v_f = 0m/s + (10m/s^2)(4s) = 40m/s$$

For part **c)**, we have one additional known variable: t= 20 s

The unknowns are: d (the distance after the truck has traveled for 20 seconds)

Equation (2) will allow us to solve this problem:

$$d = v_i t + \frac{1}{2}at^2$$

$$d = (0m/s)(20s) + \frac{1}{2}(10m/s^2)(20s)^2 = 2000m$$

Finally, we consider whether these solutions seem physically reasonable. In this simple problem, we can easily say that they do.

Skill 1.3 Interpreting algebraically and graphically relationships among position, velocity, acceleration, and time

Algebraic relationships are introduced in **Skill 1.1**.

The relationship between time, position or distance, velocity and acceleration can be understood conceptually by looking at a graphical representation of each as a function of time. Simply, the velocity is the slope of the position vs. time graph and the acceleration is the slope of the velocity vs. time graph. If you are familiar with calculus then you know that this relationship can be generalized: velocity is the first derivative and acceleration the second derivative of position.

Here are three examples:

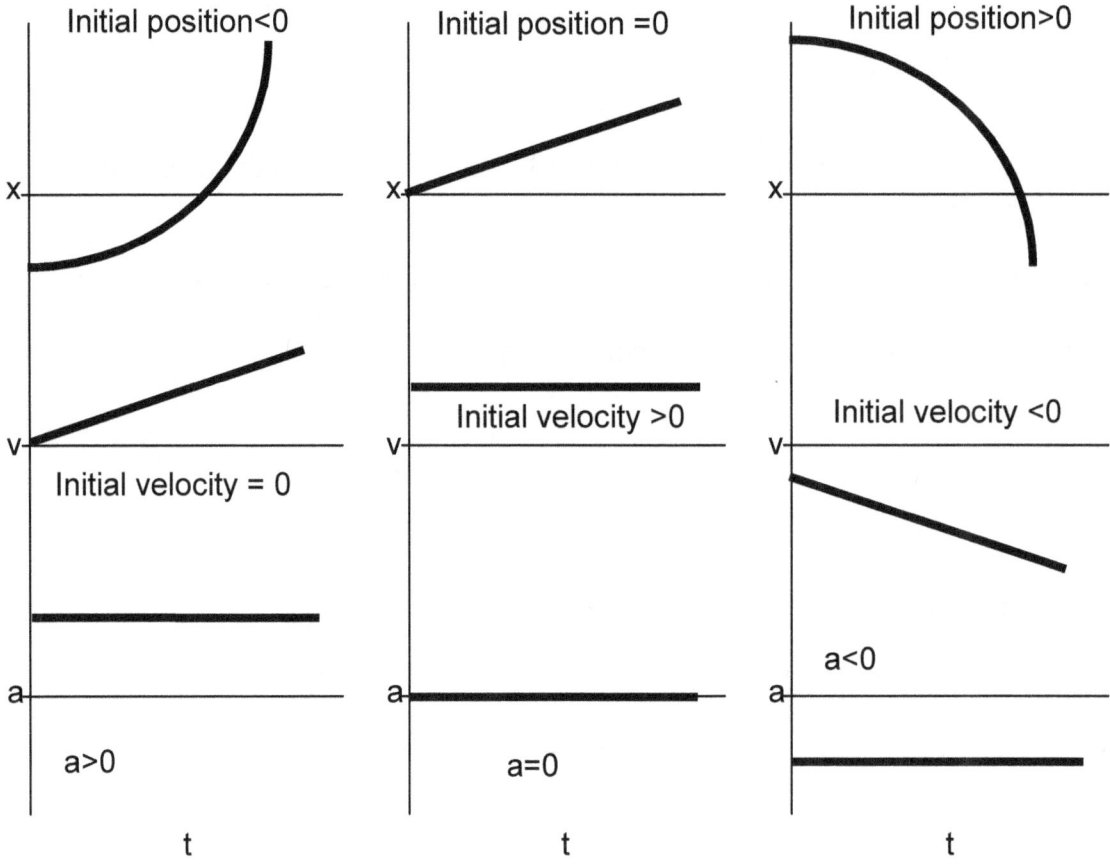

There are three things to notice:

1) In each case acceleration is constant. This isn't always the case, but a simplification for this illustration.

2) A non-zero acceleration produces a position curve that is a parabola.

3) In each case the initial velocity and position are specified separately. The acceleration curve gives the shape of the velocity curve, but not the initial value and the velocity curve gives the shape of the position curve but not the initial position.

Skill 1.4 Analyzing problems involving motion in two dimensions

In the previous section, we discussed the relationships between distance, velocity, acceleration and time and the four simple equations that relate these quantities when acceleration is constant (e.g. in cases such as gravity). In two dimensions the same relationships apply, but each dimension must be treated separately.

The most common example of an object moving in two dimensions is a projectile. A projectile is an object upon which the only force acting is gravity. Some examples:

i) An object dropped from rest.
ii) An object thrown vertically upwards at an angle
iii) A canon ball.

Once a projectile has been put in motion (say, by a canon or hand) the only force acting it is gravity, which near the surface of the earth implies it experiences $a=g=9.8 m/s^2$.

This is most easily considered with an example such as the case of a bullet shot horizontally from a standard height at the same moment that a bullet is dropped from exactly the same height. Which will hit the ground first? If we assume wind resistance is negligible, then the acceleration due to gravity is our only acceleration on either bullet and we must conclude that they will hit the ground at the same time. The horizontal motion of the bullet is not affected by the downward acceleration.

Example:
I shoot a projectile at 1000 m/s from a perfectly horizontal barrel exactly 1 m above the ground. How far does it travel before hitting the ground?

Solution:
First figure out how long it takes to hit the ground by analyzing the motion in the vertical direction. In the vertical direction, the initial velocity is zero so we can rearrange kinematic equation 2 from the previous section to give:

$t = \sqrt{\dfrac{2d}{a}}$. Since our displacement is 1 m and $a=g=9.8 m/s^2$, t=0.45 s.

Now use the time to hitting the ground from the previous calculation to calculate how far it will travel horizontally. Here the velocity is 1000m/s and there is no acceleration. So we simple multiply velocity with time to get the distance of 450m.

Motion on an arc can also be considered from the view point of the kinematic equations. As pointed out earlier, displacement, velocity and acceleration are all vector quantities, i.e. they have magnitude (the speed is the magnitude of the velocity vector) and direction. This means that if one drives in a circle at constant speed one still experiences an acceleration that changes the direction. We can define a couple of parameters for objects moving on circular paths and see how they relate to the kinematic equations.

1. Tangential speed: The tangent to a circle or arc is a line that intersects the arc at exactly one point. If you were driving in a circle and instantaneously moved the steering wheel back to straight, the line you would follow would be the tangent to the circle at the point where you moved the wheel. The tangential speed then is the instantaneous magnitude of the velocity vector as one moves around the circle.

2. Tangential acceleration: The tangential acceleration is the component of acceleration that would change the tangential speed and this can be treated as a linear acceleration if one imagines that the circular path is unrolled and made linear.

3. Centripetal acceleration: Centripetal acceleration corresponds to the constant change in the direction of the velocity vector necessary to maintain a circular path. Always acting toward the center of the circle, centripetal acceleration has a magnitude proportional to the tangential speed squared divided by the radius of the path.

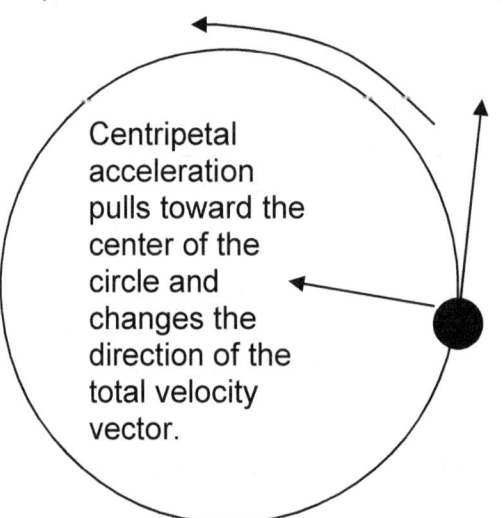

Centripetal acceleration pulls toward the center of the circle and changes the direction of the total velocity vector.

Tangential Speed= the magnitude of the velocity vector. A tangential acceleration changes the tangential speed.

Uniform circular motion describes the motion of an object as it moves in a circular path at constant speed. There are many everyday examples of this behavior though we may not recognize them if the object does not complete a full circle. For example, a car rounding a curve (that is an arc of a circle) often exhibits uniform circular motion.

The following diagram and variable definitions will help us to analyze uniform circular motion.

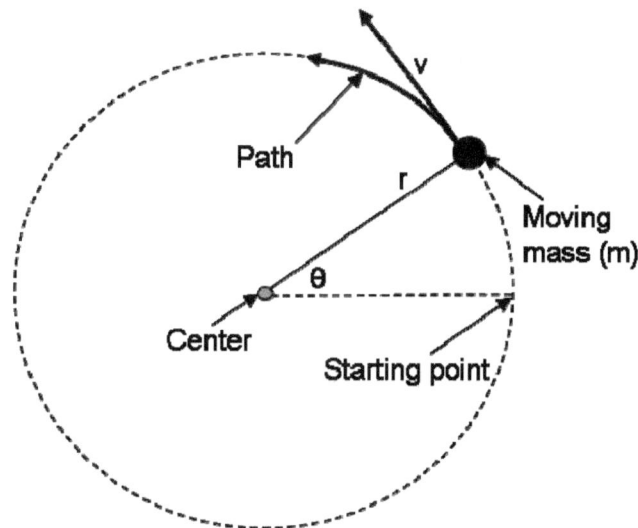

Above we see that the mass is traveling a path with constant radius (r) from some center point (x_0, y_0). By defining a variable (θ) that is a function of time (t) and is the angle between the mass's present position and original position on the circular path, we can write the following equations for the mass's position in a Cartesian plane.

$$x = r \cos(\theta) + x_0$$
$$y = r \sin(\theta) + y_0$$

Next observe that, because we are discussing uniform circular motion, the *magnitude* of the mass's velocity (v) is constant. However, the velocity's direction is always tangent to the circle and so always changing. We know that a changing velocity means that the mass must have a positive acceleration. This acceleration is directed toward the center of the circular path and is always perpendicular to the velocity, as shown below:

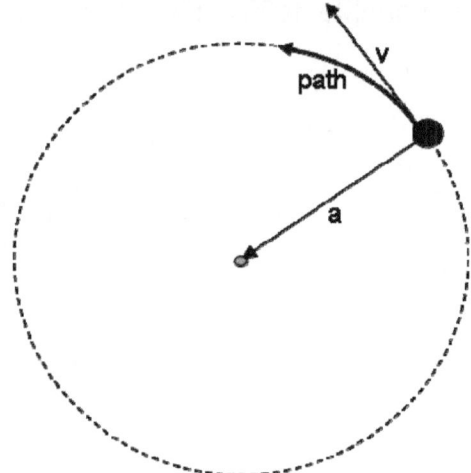

This is known as centripetal acceleration and is mathematically expressed as:

$$a = \frac{v^2}{r} = \frac{4\pi^2 r}{t^2}$$

where *t* is the period of the motion or the time taken for the mass to travel once around the circle. The force (F) experienced by the mass (m) is known as centripetal force and is always directed towards the center of the circular path. It has constant magnitude given by the following equation:

$$F = ma = m\frac{v^2}{r}$$

Skill 1.5 Analyzing properties of vectors and solving problems involving vector quantities analytically and graphically

Vector space is a collection of objects that have **magnitude** and **direction**. They may have mathematical operations, such as addition, subtraction, and scaling, applied to them. Vectors are usually displayed in boldface or with an arrow above the letter. They are usually shown in graphs or other diagrams as arrows. The length of the arrow represents the magnitude of the vector while the direction in which the arrow points shows the vector direction.

To **add two vectors** graphically, the base of the second vector is drawn from the point of the first vector as shown below with vectors **A** and **B**. The sum of the vectors is drawn as a dashed line, from the base of the first vector to the tip of the second. As illustrated, the order in which the vectors are connected is not significant as the endpoint is the same graphically whether **A** connects to **B** or **B** connects to **A**. This principle is sometimes called the parallelogram rule.

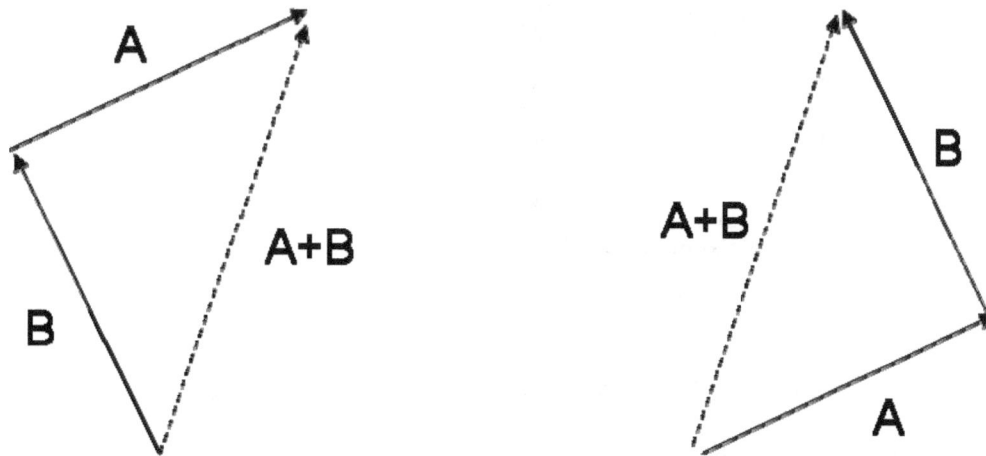

If more than two vectors are to be combined, additional vectors are simply drawn in accordingly with the sum vector connecting the base of the first to the tip of the final vector.

Subtraction of two vectors can be geometrically defined as follows. To subtract **A** from **B**, place the ends of **A** and **B** at the same point and then draw an arrow from the tip of **A** to the tip of **B**. That arrow represents the vector **B-A**, as illustrated below:

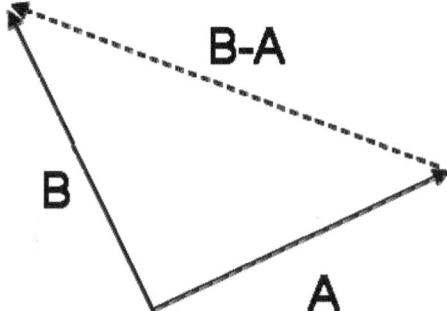

To add two vectors without drawing them, the vectors must be broken down into their orthogonal components using sine, cosine, and tangent functions. Add both x components to get the total x component of the sum vector, then add both y components to get the y component of the sum vector. Use the Pythagorean Theorem and the three trigonometric functions to the get the size and direction of the final vector.

Example: Here is a diagram showing the x and y-components of a vector D1:

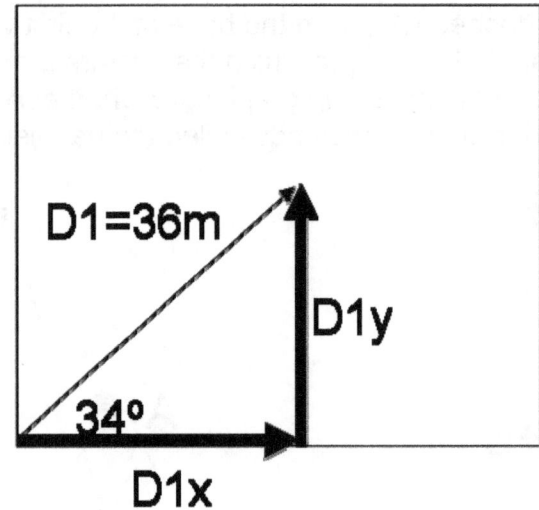

Notice that the x-component D1x is adjacent to the angle of 34 degrees.

Thus D1x=36m (cos34) =29.8m

The y-component is opposite to the angle of 34 degrees.
Thus D1y =36m (sin34) = 20.1m

A second vector D2 is broken up into its components in the diagram below using the same techniques. We find that D2y=9.0m and D2x=-18.5m.

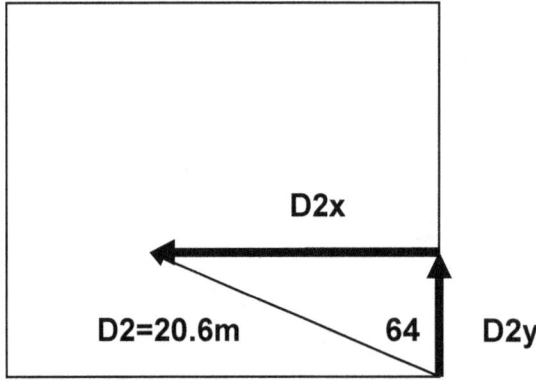

Next we add the x components and the y components to get

DTotal x =11.3 m   and   DTotal y =29.1 m

Now we have to use the Pythagorean theorem to get the total magnitude of the final vector. And the arctangent function to find the direction. As shown in the diagram below.

DTotal=31.2m

tan θ= DTotal y / DTotal x = 29.1m / 11.3 =2.6   θ=69 degrees

Vector multiplication (dot and cross product)

The dot product is also known as the scalar product. This is because the dot product of two vectors is not a vector, but a scalar (i.e., a real number without an associated direction). The definition of the dot product of the two vectors **a** and **b** is:

$$a \bullet b = \sum_{i=1}^{n} a_i b_i = a_1 b_1 + a_2 b_2 + ... + a_n b_n$$

The following is an example calculation of the dot product of two vectors:

[1 3 -5] · [4 -2 -2] = (1)(4) + (3)(-2) + (-5)(-2) = 8

Note that the product is a simple scalar quantity, not a vector. The dot product is commutative and distributive.
Unlike the dot product, the cross product does return another vector. The vector returned by the cross product is orthogonal to the two original vectors. The cross product is defined as:

a x b = n |a| |b| sin θ

where n is a unit vector perpendicular to both **a** and **b** and θ is the angle between **a** and **b**. In practice, the cross product can be calculated as explained below:

Given the orthogonal unit vectors **i**, **j**, and **k**, the vector **a** and **b** can be expressed:

$$\mathbf{a} = a_1\mathbf{i} + a_2\mathbf{j} + a_3\mathbf{k}$$
$$\mathbf{b} = b_1\mathbf{i} + b_2\mathbf{j} + b_3\mathbf{k}$$

Then we can calculate that

a x b =**i**$(a_2 b_3)$+**j**$(a_3 b_1)$+**k**$(a_1 b_2)$-**i**$(a_3 b_2)$-**j**$(a_1 b_3)$-**k**$(a_2 b_1)$

The cross product is anticommutative (that is, **a x b**= - **b x a**) and distributive over addition.

TEACHER CERTIFICATION STUDY GUIDE

COMPETENCY 2.0 UNDERSTAND NEWTON'S LAWS AND THE UNIVERSAL LAW OF GRAVITATION

Skill 2.1 Analyzing examples of Newton's laws of motion in daily life

Newton's first law of motion: "An object at rest tends to stay at rest and an object in motion tends to stay in motion with the same speed and in the same direction unless acted upon by an unbalanced force". Prior to Newton's formulation of this law, being at rest was considered the natural state of all objects, because at the earth's surface we have the force of gravity working at all times which causes nearly any object put into motion to eventually come to rest. Newton's brilliant leap was to recognize that an unbalanced force changes the motion of a body, whether that body begins at rest or at some non-zero speed.

We experience the consequences of this law everyday. For instance, the first law is why seat belts are necessary to prevent injuries. When a car stops suddenly, say by hitting a road barrier, the driver continues on forward until acted upon by a force. The seat belt provides that force and distributes the load across the whole body rather than allowing the driver to fly forward and experience the force against the steering wheel.

Example: A skateboarder is riding her skateboard down a road. The skateboard has a constant speed of 5 m/s. Then the skateboard hits a rock and stops suddenly. Since the rider has nothing to stop her when the skateboard stops, she will continue to travel at 5 m/s until she hits the ground.

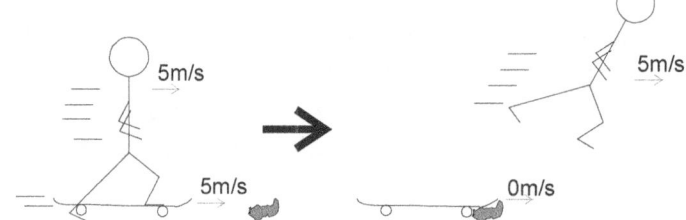

Newton's second law of motion: "The acceleration of an object as produced by a net force is directly proportional to the magnitude of the net force, in the same direction as the net force, and inversely proportional to the mass of the object". In the equation form, it is stated as $F = ma$, force equals mass times acceleration. It is important to remember that this is the net force and that forces are vector quantities. Thus if an object is acted upon by 12 forces that sum to zero, there is no acceleration. Also, this law embodies the idea of inertia as a consequence of mass. For a given force, the resulting acceleration is proportionally smaller for a more massive object because the larger object has more inertia.

Example:
A ball is dropped from a building. The mass of the ball is 2 kg. The acceleration of the object is 9.8 m/s² (gravitational acceleration). Therefore, the force acting on the ball is
$F = ma \Rightarrow F = 2 \text{ kg} \times 9.8 \text{ m/s}^2 \Rightarrow F = 19.6 \text{ N}$

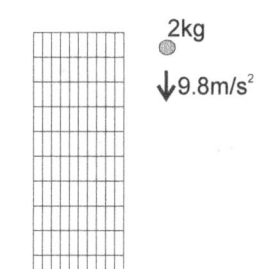

Newton's third law of motion: "For every action, there is an equal and opposite reaction". This statement means that in every interaction, there is a pair of forces acting on the two interacting objects. The size of the force on the first object equals the size of the force on the second object. The direction of the force on the first object is opposite to the direction of the force on the second object.

Example: A box is sitting on a table. The mass of the box is 4 kg. Because of the effects of gravity, the box is applying a force of 39.2 N on the table. The table does not break or shift under the force of the box. This implies that the table is applying a force of 39.2 N on the box. Note that the force that the table is applying to the box is in the opposite direction to the force that the box is applying to the table.

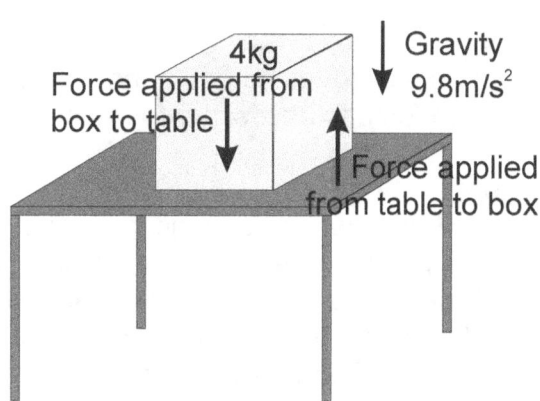

Here are a few more examples:

1. The propulsion/movement of fish through water: A fish uses its fins to push water backwards. The water pushes back on the fish. Because the force on the fish is unbalanced the fish moves forward.

2. The motion of car: A car's wheels push against the road and the road pushes back. Since the force of the road on the car is unbalanced the car moves forward.

3. Walking: When one pushes backwards on the foot with the muscles of the leg, the floor pushes back on the foot. If the forces of the leg on the foot and the floor on the foot are balanced, the foot will not move and the muscles of the body can move the other leg forward.

TEACHER CERTIFICATION STUDY GUIDE

Skill 2.2 **Applying knowledge of free-body diagrams and vector properties to solve problems**

Newton's laws of motion can be used together or separately to analyze a variety of physical situations. Simple examples are provided below and the importance of each law is highlighted.

Problem:
A 10 kg object moves across a frictionless surface at a constant velocity of 5 m/s. How much force is necessary to maintain this speed?

Solution:
Both Newton's first and second laws can help us understand this problem. First, the first law tells us that this object will continue its state of uniform speed in a straight line (since there is no force acting upon it). Additionally, the second law tells that because there is no acceleration (velocity is constant), no force is required. Thus, zero force is necessary to maintain the speed of 5 m/s.

Problem:
A car is driving down a road at a constant speed. The mass of the car is 400 kg. The force acting on the car is 4000 N and the force is in the same direction as the acceleration. What is the acceleration of the car?

Solution:

$$F = ma \Rightarrow a = \frac{F}{m} \Rightarrow a = \frac{4000N}{400kg} \Rightarrow a = 10m/s^2$$

Problem:
For the arrangement shown, find the force necessary to overcome the 500 N force pushing to the left and move the truck to the right with an acceleration of 5 m/s².

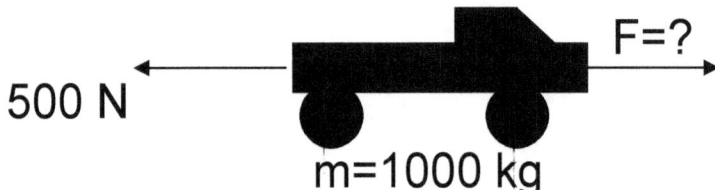

Solution:
The net force on the truck acting towards the right is F – 500N.
Using Newton's second law, F-500N = 1000kg x 5 m/s².
Solving for F, we get F = 5500 N.

PHYSICS

Problem:
An astronaut with a mass of 95 kg stands on a space station with a mass of 20,000 kg. If the astronaut is exerting 40 N of force on the space station, what is the acceleration of the space station and the astronaut?

Solution
To find the acceleration of the space station, we can simply apply Newton's second law:

$$A_s = \frac{F}{m_s} = \frac{40N}{20000kg} = 0.002 \, m/s^2$$

To find the acceleration of the astronaut, we must first apply Newton's third law to determine that the space station exerts an opposite force of -40 N on the astronaut. Here the minus sign simply denotes that the force is directed in the opposite direction. We can then calculate the acceleration, again using Newton's second law:

$$A_a = \frac{F}{m_a} = \frac{-40N}{95kg} = -0.42 \, m/s^2$$

Skill 2.3 Analyzing and solving problems involving frictional forces and coefficients

In the real world, whenever an object moves its motion is opposed by a force known as friction. How strong the frictional force is depends on numerous factors such as the roughness of the surfaces (for two objects sliding against each other) or the viscosity of the liquid an object is moving through. Most problems involving the effect of friction on motion deal with sliding friction. This is the type of friction that makes it harder to push a box across cement than across a marble floor.

When you try and push an object from rest, you must overcome the maximum **static friction** force to get it to move. Once the object is in motion, you are working against **kinetic friction** which is smaller than the static friction force previously mentioned. Sliding friction is primarily dependent on two things, the **coefficient of friction (μ)** which is primarily dependent on roughness of the surfaces involved and the amount of force pushing the two surfaces together. This force is also known as the **normal force (F_n)**, the perpendicular force between two surfaces. When an object is resting on a flat surface, the normal force is pushing opposite to the gravitational force – straight up. When the object is resting on an incline, the normal force is less (because it is only opposing that portion of the gravitational force acting perpendicularly to the object) and its direction is perpendicular to the surface of incline but at an angle from the ground. Therefore, for an object experiencing no external action, the magnitude of the normal force is either equal to or less than the magnitude of the gravitational force (F_g) acting on it. The frictional force (F_f) acts perpendicularly to the normal force, opposing the direction of the object's motion.

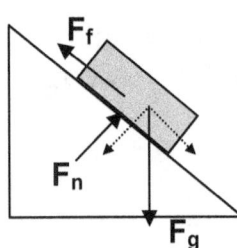
The frictional force is normally directly proportional to the normal force and, unless you are told otherwise, can be calculated as **F$_f$** = μ **F$_n$** where μ is either the coefficient of static friction or kinetic friction depending on whether the object starts at rest or in motion. In the first case, the problem is often stated as "how much force does it take to start an object moving" and the frictional force is given by **F$_f$** > μ$_s$ **F$_n$** where μ$_s$ is the coefficient of static friction. When questions are of the form "what is the magnitude of the frictional force opposing the motion of this object," the frictional force is given by **F$_f$** = μ$_k$ **F$_n$** where μ$_k$ is the coefficient of kinetic friction.

There are several important things to remember when solving problems about friction.

1. The frictional force acts in opposition to the direction of motion.

2. The frictional force is proportional, and acts perpendicular to, the normal force.

3. The normal force is perpendicular to the surface the object is lying on. If there is a force pushing the object against the surface, it will increase the normal force.

Skill 2.4   Solving problems involving gravitational forces

Newton's universal law of gravitation states that any two objects experience a force between them as the result of their masses. Specifically, the force between two masses m_1 and m_2 can be summarized as

$$F = G \frac{m_1 m_2}{r^2}$$

where G is the gravitational constant ($G = 6.672 \times 10^{-11} \, Nm^2 / kg^2$), and r is the distance between the two objects.

The weight of an object is the result of the gravitational force of the earth acting on its mass. The acceleration due to Earth's gravity on an object is 9.81 m/s². Since force equals mass * acceleration, the magnitude of the gravitational force created by the earth on an object is

$$F_{Gravity} = m_{object} \cdot 9.81 \, m/s^2$$

Important things to remember:

1. The gravitational force is proportional to the masses of the two objects, but *inversely* proportional to the *square of the distance* between the two objects.

2. When calculating the effects of the acceleration due to gravity for an object above the earth's surface, the distance above the surface is ignored because it is inconsequential compared to the radius of the earth. The constant figure of 9.81 m/s² is used instead.

Problem: Two identical 4 kg balls are floating in space, 2 meters apart.

What is the magnitude of the gravitational force they exert on each other?

Solution: $$F = G\frac{m_1 m_2}{r^2} = G\frac{4 \times 4}{2^2} = 4G = 2.67 \times 10^{-10} \, N$$

Skill 2.5 Applying knowledge of Newton's laws of motion to solve problems in one and two dimensions

Problem:
A 10 kg object moves across a frictionless surface at a constant velocity of 5 m/s. How much force is necessary to maintain this speed?

Solution:
Both Newton's first and second laws can help us understand this problem. First, the first law tells us that this object will continue its state of uniform speed in a straight line (since there is no force acting upon it). Additionally, the second law tells that because there is no acceleration (velocity is constant), no force is required. Thus, zero force is necessary to maintain the speed of 5 m/s.

Problem:
A woman is pushing an 800N box across the floor. She pushes with a force of 1000 N. The coefficient of kinetic friction is 0.50. If the box is already moving, what is the force of friction acting on the box?

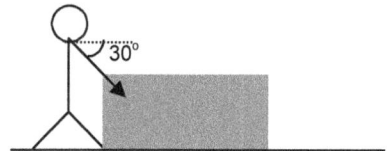

PHYSICS

Solution:
First it is necessary to solve for the normal force.
F_n= 800N + 1000N (sin 30°) = 1300N
Then, since **F_f** = μ **F_n** = 0.5*1300=650N

Problem:
An astronaut with a mass of 95 kg stands on a space station with a mass of 20,000 kg. If the astronaut is exerting 40 N of force on the space station, what is the acceleration of the space station? Of the astronaut?

Solution:
To find the acceleration of the space station, we can simply apply Newton's second law:

$$A_s = \frac{F}{m_s} = \frac{40N}{20000kg} = 0.002 \, m/s^2$$

To find the acceleration of the astronaut, we must first apply Newton's third law to determine that the space station exerts an opposite force of -40 N on the astronaut. Here the minus sign simply denotes that the force is directed in the opposite direction. We can then calculate the acceleration, again using Newton's second law:

$$A_a = \frac{F}{m_a} = \frac{-40N}{95kg} = -0.42 \, m/s^2$$

Skill 2.6 Solving problems involving torque and static equilibrium

An object is said to be in a state of equilibrium when the forces exerted upon it are balanced. That is to say, forces to the left balance the forces exerted to the right, and upward forces are balanced by downward forces. The net force acting on the object is zero and the acceleration is 0 meters per second squared. This does not necessarily mean that the object is at rest. According to Newton's first law of motion, an object at equilibrium is either at rest and remaining at rest (**static equilibrium**), or in motion and continuing in motion with the same speed and direction (**dynamic equilibrium**).

Equilibrium of forces is often used to analyze situations where objects are in static equilibrium. One can determine the weight of an object in static equilibrium or the forces necessary to hold an object at equilibrium. The following are examples of each type of problem.

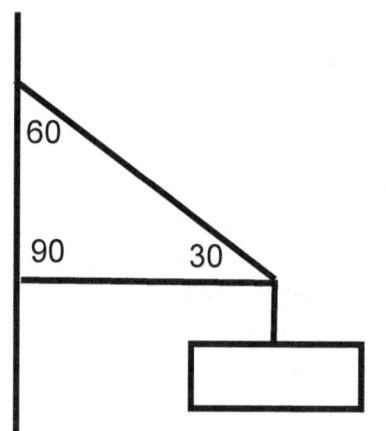

Problem: A sign hangs outside a building supported as shown in the diagram. The sign has a mass of 50 kg. Calculate the tension in the cable.

Solution: Since there is only one upward pulling cable it must balance the weight. The sign exerts a downward force of 490 N. Therefore, the cable pulls upwards with a force of 490 N. It does so at an angle of 30 degrees. To find the total tension in the cable:
$$F_{total} = 490 \text{ N} / \sin 30°$$
$$F_{total} = 980 \text{ N}$$

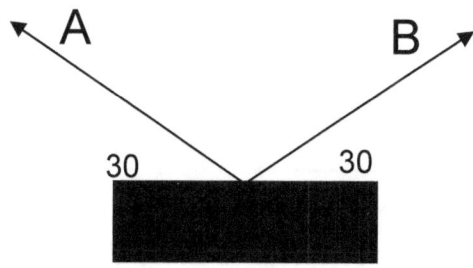

Problem: A block is held in static equilibrium by two cables. Suppose the tension in cables A and B are measured to be 50 Newtons each. The angle formed by each cable with the horizontal is 30 degrees. Calculate the weight of the block.

Solution: We know that the upward pull of the cable must balance the downward force of the weight of the block and the right pulling forces must balance the left pulling forces. Using trigonometry we know that the y component of each cable can be calculated as:
$$F_y = 50 \text{ N} \sin 30°$$
$$F_y = 25 \text{ N}$$

Since there are two cables supplying an upward force of 25 N each, the overall downward force supplied by the block must be 50 N.

Torque is rotational motion about an axis. It is defined as $\tau = L \times F$, where L is the lever arm. The length of the lever arm is calculated by measuring the perpendicular line drawn from the line of force to the axis of rotation.

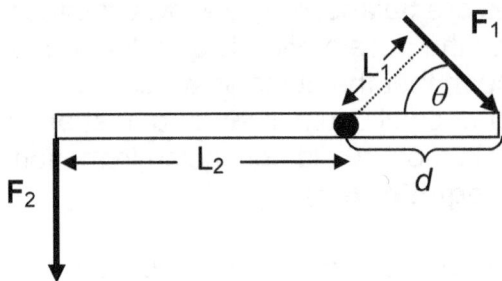

By convention, torques that act in a clockwise direction are considered negative and those in a counterclockwise direction are considered positive. In order for an object to be in equilibrium, the sum of the torques acting on it must be zero.

The equation that would put the above figure in equilibrium is $F_1 L_1 = -F_2 L_2$ (please note that in this case $L_1 = d\sin\theta$).

For an object to be in equilibrium the forces acting on it must be balanced. This applies to linear as well as rotational forces known as moments or torques. In the two dimensional example below, torque can only be applied in two directions; clockwise and counter clockwise. The convention is that positive rotation is counter clockwise and negative is clockwise. For the object to be in equilibrium, the sum of the applied torques must be zero, in addition to the sum of all forces being zero.

Let us consider a horizontal bar at equilibrium so that the bar experiences neither rotation nor translation. We can define rotation by choosing any point along the bar and labeling it A for axis.

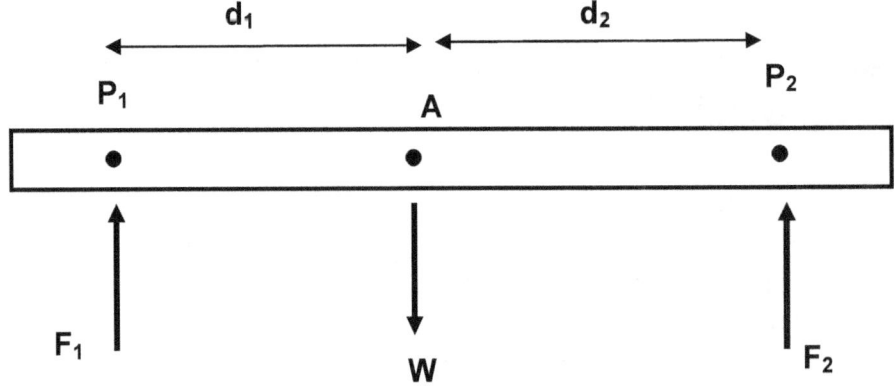

The bar experiences two point forces at either end, labeled F_1 and F_2. The torque applied to the bar by each of these forces is given by multiplying the force by the moment arm, the distance between the point where the force is applied and the axis. In the case of F_1 the torque is as follows (negative since rotation is clockwise):

$$\tau_1 = -F_1 d_1$$

The torque applied by F2 is given by (positive since rotation is counterclockwise):

$$\tau_2 = F_2 d_2$$

The other force acting on the bar is the force of gravity, or the weight of the bar. This is not a point force, but rather acts at all points along the bar. However, we can consider that the weight acts in the center of the bar, at a point called the center of mass. In the case of this example, we are taking the axis to be located at the center of mass. Since the axis is located at the center of mass, the torque exerted on the bar due to its weight is zero.

Suppose $F_1=2$ N, $d_1=0.4$ m, and $d_2=0.5$ m. Let us calculate F_2. We will use our knowledge that the sum of all torques must equal zero when that object is at equilibrium.

$$\tau_1 + \tau_2 = 0$$

$$-F_1 d_1 + F_2 d_2 = 0$$
$$(-2 \text{ N} \times 0.4 \text{ m}) + (F_2 \times 0.5 \text{ m}) = 0$$
$$-0.8 + 0.5 F_2 = 0$$
$$F_2 = 1.6 \text{ N}$$

It is also possible to calculate the weight of the bar since we know that the sum of all forces must be zero. Since F1 and F2 act up but weight acts down we have:

$$2 \text{ N} + 1.6 \text{ N} - W = 0$$
$$W = 3.6 \text{ N}$$

Problem:
Some children are playing with the spinner below, when one young boy decides to pull on the spinner arrow in the direction indicated by **F₁**. How much torque does he apply to the spinner arrow?

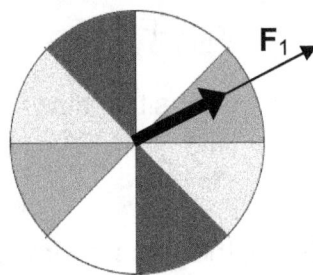

$\tau = L \times F$, but L=0 because the line of force goes directly through the axis of rotation (i.e. the perpendicular distance from the line of force to the pivot point is 0). Therefore the boy applies no torque to the spinner.

Problem: In the system diagrammed below, find out what the magnitude of F_1 must be in order to keep the system in equilibrium.

Solution: For an object to be in equilibrium the forces acting on it must be balanced. This applies to linear as well as rotational forces known as moments or torques.

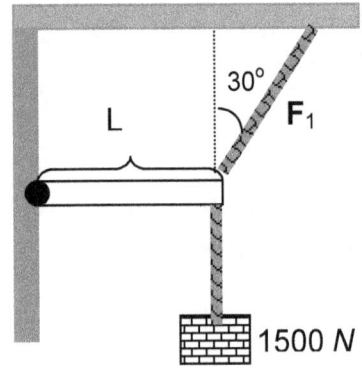

$$F_1 L = -F_2 L$$
$$F_{1x} L + F_{1y} L = -(-1500L) \ldots \text{but } F_{1x} = 0 *$$
$$F_{1y} L = 1500 L$$
$$F_1 \cos 30 L = 1500 L$$
$$(0.866) F_1 = 1500$$
$$F_1 = 1732 N$$

*the effect of F_{1x}=0 because that portion of the force goes right through the pivot and causes no torque.

Things to remember:

1. When writing the equation for a body at equilibrium, the point chosen for the axis of rotation is arbitrary.
2. The center of gravity is the point in an object where its weight can be considered to act for the purpose of calculating torque.
3. The lever arm, or moment arm, of a force is calculated as the **perpendicular** distance from the line of force to the pivot point/axis of rotation.
4. Counterclockwise torques are considered positive while clockwise torques are considered negative.

COMPETENCY 3.0 UNDERSTAND THE CONSERVATION OF ENERGY

Skill 3.1 Calculating the kinetic and potential energy of mechanical systems

Energy can be defined, in relation to work, as the ability of an object to do work. As such, it is measured in the same units as work, usually Joules. Most problems relating work to energy are looking at two specific kinds of energy. The first, kinetic energy, is the energy of motion. The heavier an object is and the faster it is going, the more energy it has resulting in a greater capacity for work. The equation for kinetic energy is: $KE = \frac{1}{2}mv^2$.

Problem:
A 1500 kg car is moving at 60m/s down a highway when it crashes into a 3000kg truck. In the moment before impact, how much kinetic energy does the car have?

Solution:
$KE = \frac{1}{2}mv^2 = \frac{1}{2} \cdot 1500 \cdot 60^2 = 2.7 \times 10^6 \, J$

The other form of energy frequently discussed in relationship to work is gravitational potential energy, or potential energy, the energy of position. Potential energy is calculated as $PE = mgh$ where h is the distance the object is capable of falling.

Problem:
Which has more potential energy, a 2 kg box held 5 m above the ground or a 10 kg box held 1 m above the ground?

Solution:

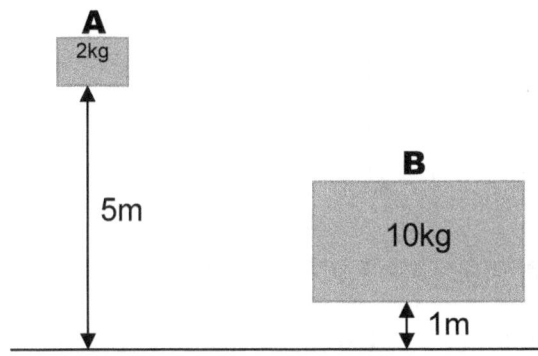

$PE_A = mgh = 2 \cdot g \cdot 5 = 10g$
$PE_B = mgh = 10 \cdot g \cdot 1 = 10g$
$PE_A = PE_B$

Skill 3.2 Applying knowledge of the law of conservation of energy and the work-energy theorem to solve problems involving conservative and nonconservative forces

According to the concept of conservation of energy, the energy in an isolated system remains the same although it may change in form. For instance, potential energy can become kinetic energy and kinetic energy, depending on the system, can become thermal or heat energy. Solving energy conservation problems depends on knowing the types of energy one is dealing with in a particular situation and assuming that the sum of all the different types of energy remains constant. Below we will discuss several different examples.

Example:
A rollercoaster at the top of a hill has a certain potential energy that will allow it to travel down the track at a speed based on its potential energy and friction with the track itself. At the bottom of the hill, when it has reached a stop, its potential energy is zero and all of the energy has been transferred from potential energy to kinetic energy (movement) and thermal energy (heat derived from friction). The equation below describes the relationship between potential energy and other forms of energy in this case:

Potential Energy = Kinetic energy(movement) + heat energy(friction)

Problem:
A skier travels down a ski slope with negligible friction. He begins at 100 meters in height, drops to a much lower level and ends at 90 meters in height. What is the skier's velocity at the 90 meter height?

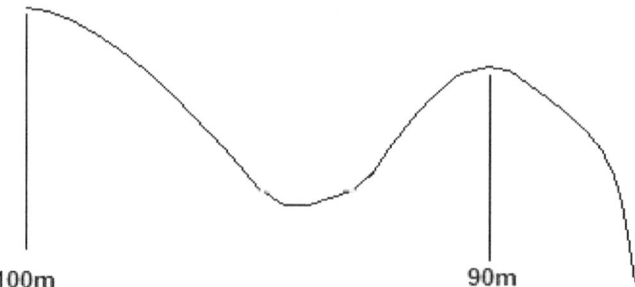

100m 90m

The skier's initial energy is only potential energy and is given by $mgh = 100mg$ where m is the mass of the skier.
The skier's final energy is the sum of his potential and kinetic energies and is given by $mgh + 1/2mv^2 = 90mg + 1/2mv^2$ where v is the skier's velocity.
Using the principle of conservation of energy we know that the initial and final energies must be equal. Hence
$$100mg = 90mg + 1/2mv^2$$

Thus $1/2mv^2 = 10mg$; $1/2v^2 = 10g$; $v = 14 m/s$ since ($g = 9.81 m/s^2$)

Problem:
d. A pebble weighing 10 grams is placed in a massless frictionless sling shot, with a spring constant of 200N/m, that is stretched back 0.5 meters. What is the total energy of the system before the pebble is released? What is the final height of the pebble if it is shot straight up and the effects of air resistance are negligible?

Solution:
If the initial height of the pebble is h = 0, total energy is given by
E = ½ mv² + mgh + ½ kx² = 0 + 0 + 0.5(200)(0.5)² = 25 Joules

At its final height, the velocity of the pebble will be zero. Since
E = ½ mv² + mgh + ½ kx², from the principle of conservation of energy
25 Joules = 0 + 0.010kg (9.81)h + 0 and h = 255 m

The work-energy theorem states that **the amount of work done on an object is equal to its change in mechanical energy (kinetic or potential energy)**. Specifically, in systems where multiple forces are at work, the energy change of the system is the work done by the *net* force on the object. Problems dealing with the work-energy theorem may look at changes in kinetic energy, changes in potential energy, or some combination of the two. It is also important to remember that only external forces can cause changes in an object's total amount of mechanical energy. Internal forces, such as spring force or gravity, only lead to conversions between kinetic and potential energy rather than changes in the total level of mechanical energy.

Problem:
A woman driving a 2000 kg car along a level road at 30 m/s takes her foot off the gas to see how far her car will roll before it slows to a stop. She discovers that it takes 150m. What is the average force of friction acting on the car?

Solution:

$W = \Delta KE$

$f \cdot s \cos \theta = \frac{1}{2} m v^2_{final} - \frac{1}{2} m v^2_{initial}$

$f \cdot 150 \cdot (-1) = \frac{1}{2} \cdot 2000 \cdot 0^2 - \frac{1}{2} \cdot 2000 \cdot 30^2$

$-150 f = -900000$

$f = 6000 N$

According to the work-energy theorem, the amount of work done on the car is equal to the change in its mechanical energy which in this case is its change in kinetic energy. Since the only force acting on the car is friction, all the work can be attributed to the frictional force.

It is important to realize, for this problem, that the force of friction is against the direction of motion, and thus cosθ = -1.

PHYSICS 25

Problem:
A 20kg child lifts his .5 kg ball off the floor to put it away on his bookshelf 1.5 meters above the ground. How much work has he done?

Solution: The work done is equal to the change in potential energy of the book.

$$W = \Delta PE$$
$$W = mgh_{final} - mgh_{initial}$$
$$W = mg(h_{final} - h_{initial}) = .5 \cdot 9.8 \cdot 1.5 = 7.35 J$$

The principle of conservation of energy states that an isolated system maintains a constant total amount of energy despite the fact that the energy may change forms. To put it another way, energy cannot be created or destroyed but can be changed from one form to another. For example, friction can turn kinetic energy into thermal energy. Other forms of energy include electrical energy, chemical energy, and mechanical energy.

A **conservative** force is one that conserves mechanical energy (kinetic + potential energy), i.e. there is no change in mechanical energy when a conservative force acts on an object. Consider a mass on a spring on a frictionless surface. This is a closed loop system. If conservative forces alone act on the mass during each cycle, the velocity of the mass at the beginning and the end of the cycle must be the same for the mechanical energy to have been conserved. In this way, the force has done no work. At any point in the cycle of motion, the total mechanical energy of the system remains constant even though the energy moves back and forth between kinetic and potential forms. If work is done on the mass, then the forces acting on the mass are **nonconservative**. In a real system there will be some dissipative forces that will convert some of the mechanical energy to thermal energy. Conservative forces are independent of path the object takes, while nonconservative forces are path dependent.

Gravity is a conservative force. This can be illustrated by imagining an object tossed into the air. On the upward journey the work done by gravity is the negative product of mass, acceleration, and height. On the downward journey, the work done by gravity is the positive value of this amount. Thus for the total loop the work is zero.

Friction is a nonconservative force. If a box is pushed along a rough surface from one side of the room to the other and back, friction opposes the movement in both directions; so the work done by friction cannot be equal to zero.

This example also helps illustrate how nonconservative forces are path dependent. More work is done by friction if the path is tortuous rather than straight, even if the start and end points are the same. Let's try an example with a small box of mass 5 kg. The box moves in a circle 2 meters in diameter. The coefficient of kinetic friction between the box and the surface it rests on is 0.2. How much work is done by friction during one revolution?

The force exerted by friction is calculated by

$$F_k = \mu_k F_n = (0.2)(5 \text{ kg})(9.8 \text{ m/s}^2) = 9.8 \text{ N}$$

The force opposes the movement of the box during the entire distance of one revolution, or approximately 6.3 meters ($2\pi r$).

The total work done by friction is

$$W = F \times \cos\theta = (9.8 \text{ N})(6.3 \text{ m})(\cos 180) = -61.7 \text{ Joules}$$

As expected, the work is not zero since friction is not a conservative force. Since it does negative work on an object, it reduces the mechanical energy of the object and is a **dissipative** force.

Skill 3.3 Analyzing mechanical systems in terms of work, power, and energy

In physics, work is defined as force times distance $W = F \cdot s$. Work is a scalar quantity, it does not have direction, and it is usually measured in Joules ($N \cdot m$). It is important to remember, when doing calculations about work, that the only part of the force that contributes to the work is the part that acts in the direction of the displacement. Therefore, sometimes the equation is written as $W = F \cdot s \cos\theta$, where θ is the angle between the force and the displacement.

Problem:
A man uses 6N of force to pull a 10kg block, as shown below, over a distance of 3 m. How much work did he do?

Solution:

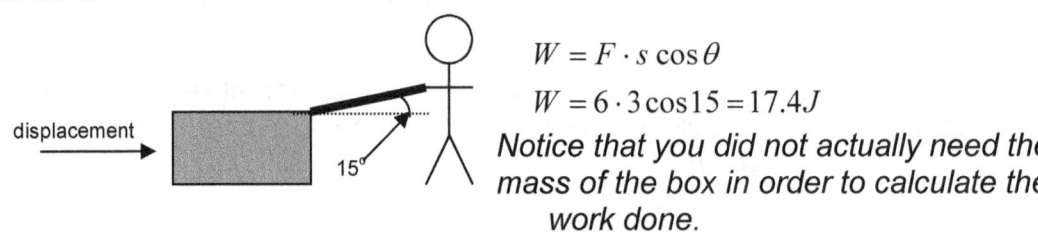

$W = F \cdot s \cos\theta$
$W = 6 \cdot 3 \cos 15 = 17.4 J$
Notice that you did not actually need the mass of the box in order to calculate the work done.

The power expended by a system can be defined as either the rate at which work is done by the system or the rate at which energy is transferred from it. There are many different measurements for power, but the one most commonly seen in physics problems is the Watt which is measured in Joules per second. Another commonly discussed unit of power is horsepower, and 1hp=746 W.

The **average power** of a system is defined as the work done by the system divided by the total change in time:

$\overline{P} = \dfrac{W}{\Delta t}$ ⇒ Where $\overline{P}$ = average power, W = work and Δt = change in time

The average power can also be written in terms of energy transfer ⇒ $\overline{P} = \dfrac{\Delta E}{\Delta t}$ and used the same way that the equation for work is used.

Problem:
A woman standing in her 4th story apartment raises a 10kg box of groceries from the ground using a rope. She is pulling at a constant rate, and it takes her 5 seconds to raise the box one meter. How much power is she using to raise the box?

Solution:

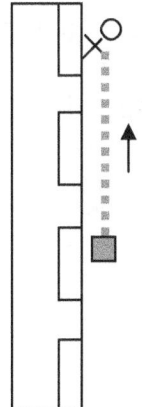

$$P = W/t$$
$$P = \frac{F \cdot s}{t} = \frac{mgh}{t} = \frac{10*9.8*1}{5} = 19.6W$$

Notice that because she is pulling at a constant rate, you don't need to know the actual distance she has raised the box. 2 meters in 10 seconds would give you the same result as 5 meters in 25 seconds.

Instantaneous power is the power measured or calculated in an instant of time. Since instantaneous power is the rate of work done when Δt is approaching 0s, the power is then written in derivate form:

$P = \frac{dW}{dt} \Rightarrow$ Where P = average power, dW = work and dt = change in time.

Since $W = Fs\cos\phi$, for a constant force the above equation can be written as:

$$P = \frac{dW}{dt} \Rightarrow P = \frac{d(Fs\cos\phi)}{dt} \Rightarrow P = \frac{(F\cos\phi)ds}{dt} \Rightarrow P = F\cos\phi\left(\frac{ds}{dt}\right) \Rightarrow P = Fv\cos\phi$$

where v is the velocity of the object.

COMPETENCY 4.0 UNDERSTAND MOMENTUM AND ITS CONSERVATION

Skill 4.1 Applying knowledge of the concept of impulse and the conservation of momentum to solve problems in one and two dimensions

The **impulse-momentum theorem** states that any impulse acting on a system changes the momentum of that system. When considering the impulse-momentum theorem, there are several factors that need to be taken into account. The first factor is that momentum is a vector quantity $p = m \cdot v$. It has both magnitude and direction. Therefore, any action that causes either the speed or the direction of an object to change causes a change in its momentum. An impulse is defined as a force acting over a period of time (integral of force over time), and any impulse acting on the system is equivalent to a change in its momentum, as you can see from the equations below:

$$F = m \cdot a \rightarrow F = m \cdot \frac{\Delta v}{t} \rightarrow F \cdot t = m \cdot \Delta v$$

i.e. Forces acting over time cause a change in momentum.

Sample Problems:

1. A 1 kg ball is rolled towards a wall at 4 m/s. It hits the wall, and bounces back off the wall at 3 m/s.

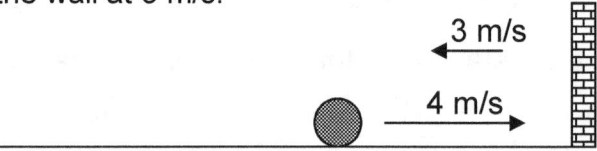

   a. What is the change in velocity?

   The velocity goes from +4m/s to –3m/s, a net change of -7m/s.

   b. At what point does the impulse occur?

   The impulse occurs when the ball hits the wall.

2. A 30kg woman is in a car accident. She was driving at 50m/s when she had to hit the brakes to avoid hitting the car in front of her.

   a. The automatic tensioning device in her seatbelt slows her down to a stop over a period of one half second. How much force does it apply?

   $$F = m \cdot \frac{\Delta v}{t} \rightarrow F = 30 \cdot \frac{50}{.5} = 3000N$$

   b. If she hadn't been wearing a seatbelt, the windshield would have stopped her in .001 seconds. How much force would have been applied there?

   $$F = m \cdot \frac{\Delta v}{t} \rightarrow F = 30 \cdot \frac{50}{.001} = 1500000N$$

The **law of conservation of momentum** states that the total momentum of an *isolated system* (not affected by external forces and not having internal dissipative forces) always remains the same. For instance, in any collision between two objects in an isolated system, the total momentum of the two objects after the collision will be the same as the total momentum of the two objects before the collision. In other words, any momentum lost by one of the objects is gained by the other.

Skill 4.2 Applying knowledge of the concepts of energy and momentum to analyze elastic and inelastic collisions

A collision may be **elastic** or **inelastic**. In a totally elastic collision, the kinetic energy is conserved along with the momentum. In a totally inelastic collision, on the other hand, the kinetic energy associated with the center of mass remains unchanged but the kinetic energy relative to the center of mass is lost. An example of a totally inelastic collision is one in which the bodies stick to each other and move together after the collision. Most collisions are neither perfectly elastic nor perfectly inelastic and only a portion of the kinetic energy relative to the center of mass is lost.

Example: Inelastic Collision

Imagine two carts rolling towards each other as in the diagram below

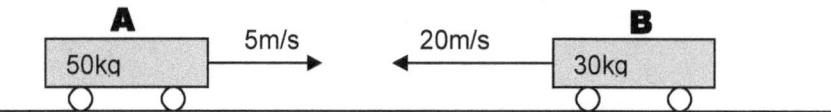

Before the collision, cart **A** has 250 kg m/s of momentum, and cart **B** has –600 kg m/s of momentum. In other words, the system has a total momentum of –350 kg m/s of momentum.

After the collision, the two carts stick to each other, and continue moving. How do we determine how fast, and in what direction, they go?

We know that the new mass of the cart is 80kg, and that the total momentum of the system is –350 kg m/s. Therefore, the velocity of the two carts stuck together must be $\frac{-350}{80} = -4.375 \, m/s$

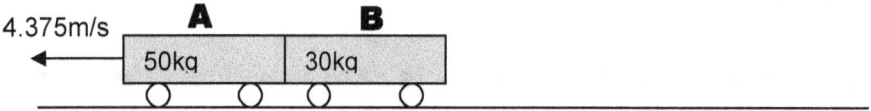

For a multidimensional example of an *elastic* collision, see **Skill 4.3**.

Skill 4.3 Applying knowledge of vectors to solve momentum problems

Conservation of momentum works the same way in two dimensions (see **Skill 4.2** for introduction to momentum problems), the only change is that you need to use vector math to determine the total momentum and any changes, instead of simple addition.

Example: Elastic Collision

Imagine a pool table like the one below. Both balls are 0.5 kg in mass.

Before the collision, the white ball is moving with the velocity indicated by the solid line and the black ball is at rest.
After the collision the black ball is moving with the velocity indicated by the dashed line (a 135° angle from the direction of the white ball).

With what speed, and in what direction, is the white ball moving after the collision?

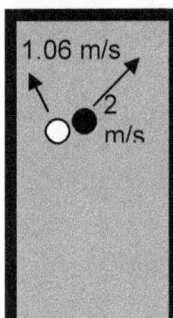

$p_{white/before} = .5 \cdot (0,3) = (0,1.5) \quad p_{black/before} = 0 \quad p_{total/before} = (0,1.5)$

$p_{black/after} = .5 \cdot (2\cos 45, 2\sin 45) = (0.71, 0.71)$

$p_{white/after} = (-0.71, 0.79)$

i.e. the white ball has a velocity of

$v = \sqrt{(-.71)^2 + (0.79)^2} = 1.06 m/s$, and is moving at an angle

of $\theta = \tan^{-1}\left(\dfrac{0.79}{-0.71}\right) = -48°$ from the horizontal

Skill 4.4 Analyzing examples of conservation of angular momentum in everyday life

The major concepts of linear motion are duplicated in rotational motion with linear displacement replaced by the angle of displacement.

Angular velocity ω = angular displacement / time

Also, the linear velocity v of a rolling object can be written as $v = r\omega$.

One important difference in the equations relates to the use of mass in rotational systems. In rotational problems, not only is the mass of an object important but also its location. In order to include the spatial distribution of the mass of the object, a term called **moment of inertia** is used, $I = m_1 r_1^2 + m_2 r_2^2 + \cdots + m_n r_n^2$

Now let's examine some physical examples of angular momentum, taking this opportunity to also perform some additional mathematical analysis of these phenomena.

Example:

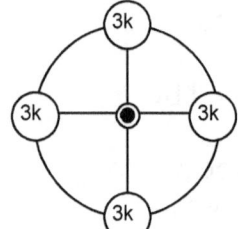

If the radius of the wheel on the left is 0.75m, what is its moment of inertia?

$$I = 3 \cdot 0.75^2 + 3 \cdot 0.75^2 + 3 \cdot 0.75^2 + 3 \cdot 0.75^2 = 6.75$$

Note: $I_{Sphere} = \frac{2}{5}mr^2$, $I_{Hoop/Ring} = mr^2$, $I_{disk} = \frac{1}{2}mr^2$

A related concept is the radius of gyration (*k*), which is the average distance of mass from the center of an object. $k_{Sphere} = \sqrt{\frac{2}{5}}r$, $k_{Hoop/Ring} = r$, $k_{disk} = \frac{r}{\sqrt{2}}$. As you can see $I = mk^2$

Angular momentum (*L*), and rotational kinetic energy (KE$_r$), are therefore defined as follows: $L = I\omega$, $KE_r = \frac{1}{2}I\omega^2$

Unless a net torque acts on a system, the angular momentum remains constant in both magnitude and direction. This can be used to solve many different types of problems including ones involving satellite motion.

Example:
A planet of mass *m* is circling a star in an orbit like the one below. If its velocity at point A is 60,000m/s, and $r_B = 8\, r_A$, what is its velocity at point B?

$$I_B \omega_B = I_A \omega_A$$
$$mr_B^2 \omega_B = mr_A^2 \omega_A$$
$$r_B^2 \omega_B = r_A^2 \omega_A$$
$$r_B^2 \frac{v_B}{r_B} = r_A^2 \frac{v_A}{r_A}$$
$$r_B v_B = r_A v_A$$
$$8 r_A v_B = r_A v_A$$
$$v_B = \frac{v_A}{8} = 7500 m/s$$

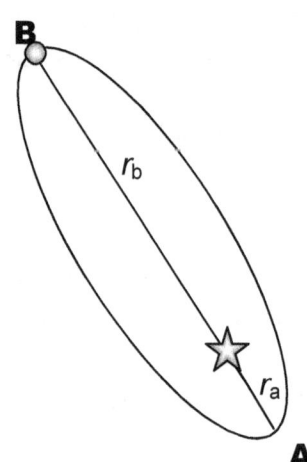

Angular momentum (L), and **rotational kinetic energy (KE$_r$)**, are therefore defined as follows: $L = I\omega$, $KE_r = \frac{1}{2}I\omega^2$

As with all systems, energy is conserved unless the system is acted on by an external force. This can be used to solve problems such as the one below.

Example:

A uniform ball of radius r and mass m starts from rest and rolls down a frictionless incline of height h. When the ball reaches the ground, how fast is it going?

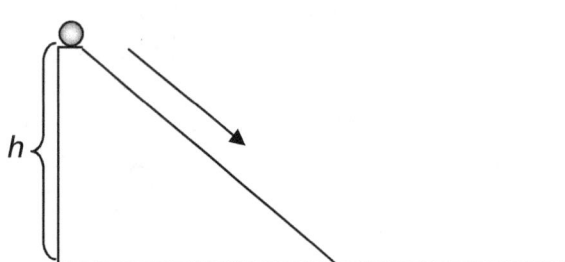

$$PE_{initial} + KE_{rotational/initial} + KE_{linear/initial} = PE_{final} + KE_{rotational/final} + KE_{linear/final}$$

$$mgh + 0 + 0 = 0 + \frac{1}{2}I\omega_{final}^2 + \frac{1}{2}mv_{final}^2 \rightarrow mgh = \frac{1}{2} \cdot \frac{2}{5}mr^2\omega_{final}^2 + \frac{1}{2}mv_{final}^2$$

$$mgh = \frac{1}{5}mr^2(\frac{v_{final}}{r})^2 + \frac{1}{2}mv_{final}^2 \rightarrow mgh = \frac{1}{5}mv_{final}^2 + \frac{1}{2}mv_{final}^2$$

$$gh = \frac{7}{10}v_{final}^2 \rightarrow v_{final} = \sqrt{\frac{10}{7}gh}$$

SUBAREA II. WAVES AND THERMAL ENERGY

COMPETENCY 5.0 UNDERSTAND THE CHARACTERISTICS OF MECHANICAL WAVES

Skill 5.1 Analyzing models of harmonic motion

Simple harmonic (sinusoidal) motion involves a cyclical exchange of kinetic energy and potential energy as observed in a simple pendulum or a mass on a spring. The relationships among the various parameters of a system displaying simple harmonic motion depends on the type of system being examined. Once the displacement is known, the velocity and acceleration of the object undergoing harmonic motion can be calculated by calculating the first derivative with respect to time (for velocity) or the second derivative with respect to time (for acceleration).

The examples of the mass on a spring and the simple pendulum represent harmonic motion in one dimension and two dimensions, respectively. Each can be treated in a similar manner, although the details vary slightly. It can be shown, however, that the pendulum acts just like the mass on a spring when the displacement angle is small.

Linear oscillator (mass on a spring)
Given the frequency of oscillation f for a mass on a spring (a linear oscillator), along with the concomitant period T = 1/f, the displacement of the mass undergoing harmonic (sinusoidal) motion can be written as follows.

$$x(t) = x_{max} \cos(2\pi f t + \phi) = x_{max} \cos(\omega t + \phi)$$

Alternatively, 2πf can be written as the angular frequency ω. The coefficient x_{max} is the maximum displacement of the mass, and the term Φ is a phase constant that determines the position of the mass at time t = 0. If x(t) is differentiated once with respect to time, the velocity of the mass is revealed.

$$v(t) = \frac{\partial x(t)}{\partial t} = -\omega x_{max} \sin(\omega t + \phi)$$

Comparing the expressions for displacement and velocity, we can see that the velocity is maximized when the displacement is zero (all kinetic energy), and the displacement is maximized when the velocity is zero (all potential energy); that is, the displacement and velocity are 90^0 out of phase. The acceleration can be calculated by differentiating v(t) with respect to time.

$$a(t) = \frac{\partial v(t)}{\partial t} = -\omega^2 x_{max} \cos(\omega t + \phi) = -\omega^2 x(t)$$

The acceleration, as shown above, is in phase with the displacement. Applying Newton's second law of motion leads to Hooke's law, which relates the restoring force on the mass to the displacement x(t).

$$F = ma = -m\omega^2 x(t)$$

This equation may be expressed in terms of the so-called spring constant k, which is defined as $m\omega^2$.

$$F = ma = -k\,x(t)$$

Simple pendulum
The simple pendulum model provides a reasonably accurate representation of pendulum motion, especially in the case of small angular amplitude.

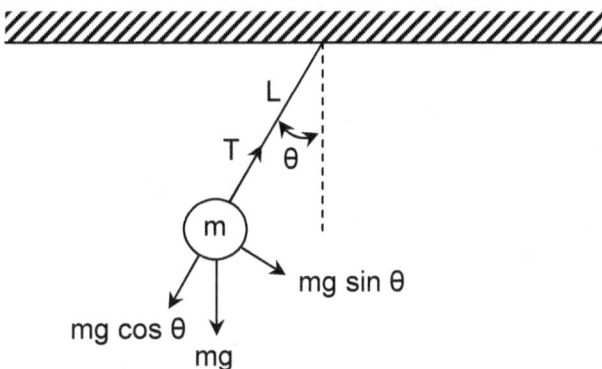

The restoring force F in pendulum motion is expressed as a component of the gravitational force mg perpendicular to the length of the string and is given by

$$F = -mg\sin\theta$$

The negative sign results from the force having a direction opposite to the displacement. The tension T on the string and the portion of the gravitational force in the opposite direction of T cancel one another.

In the case of small θ, sin θ is approximately equal to θ. The arc length traveled by the pendulum, s, is equal to the product of the length L of the string and the angle θ. Thus, the following expression can be derived.

$$F \approx -mg\theta = -mg\frac{s}{L} = -\frac{mg}{L}s$$

In the above equation, the force is shown to be of the same form as the linear harmonic oscillator, having, in this case, a "spring constant" of mg/L. As a result, the expressions found for the displacement, velocity and acceleration in the case of the linear oscillator can also be used here (in the case of small θ), where the frequency ω is replaced as follows.

$$\omega = \sqrt{\frac{g}{L}}$$

Skill 5.2   Analyzing the production and propagation of sound waves

Sound waves are mechanical waves that can travel through various kinds of media, solid, liquid and gas. They are **longitudinal waves** transmitted as variations in the pressure of the surrounding medium. These variations in pressure form waves that are termed sound waves or acoustic waves. When the particles of the medium are drawn close together it is called **compression**. When the particles of the medium are spread apart it is known as **rarefaction**.

Sound waves have different characteristics in different materials. Boundaries between different media can result in partial or total reflection of sound waves. Thus, the phenomenon of echo takes place when a sound wave strikes a material of differing characteristics than the surrounding medium. The reflection of a voice off the walls of a room is a particular example.

The frequency of the harmonic acoustic wave, ω, determines the **pitch** of the sound in the same manner that the frequency determines the color of an electromagnetic harmonic wave. **A high pitch sound corresponds to a high frequency sound wave** and a low pitch sound corresponds to a low frequency sound wave. Similarly, the amplitude of the pressure determines the **loudness** or just as the amplitude of the electric (or magnetic) field **E** determines the brightness or intensity of a color. **The intensity of a sound wave is proportional to the square of its amplitude.**

The decibel scale is used to measure sound intensity. It originated in a unit known as the bel which is defined as the reduction in audio level over 1 mile of a telephone cable. Since the bel describes such a large variation in sound, it became more common to use the decibel, which is equal to 0.1 bel. A decibel value is related to the intensity of a sound by the following equation:

$$X_{dB} = 10\log_{10}\left(\frac{X}{X_0}\right)$$

Where X_{dB} is the value of the sound in decibels
    X is the intensity of the sound
    X_0 is a reference value with the same units as X. X_0 is commonly taken to be the threshold of hearing at $10^{-12} W/m^2$.

It is important to note the logarithmic nature of the decibel scale and what this means for the relative intensity of sounds. The perception of the intensity of sound increases logarithmically, not linearly. Thus, an increase of 10 dB corresponds to an increase by one order of magnitude. For example, a sound that is 20 dB is not twice as loud as sound that is 10 dB; rather, it is 10 times as loud. A sound that is 30 dB will the 100 times as loud as the 10 dB sound.

Finally, let's equate the decibel scale with some familiar noises. Below are the decibel values of some common sounds.

Whispering voice: 20 dB
Quiet office: 60 dB
Traffic: 70 dB
Cheering football stadium: 110 dB
Jet engine (100 feet away): 150 dB
Space shuttle liftoff (100 feet away): 190 dB

Factors that affect the speed of sound in different media:

Mechanical waves rely on the local oscillation of each atom in a medium, but the material itself does not move; only the energy is transferred from atom to atom. Therefore the material through which the mechanical wave is traveling greatly affects the wave's propagation and speed. In particular, a material's elastic constant and density affect the speed at which a wave travels through it. Both of these properties of a medium can predict the extent to which the atoms will vibrate and pass along the energy of the wave. The general relationship between these properties and the speed of a wave in a solid is given by the following equation:

$$V = \sqrt{\frac{C_{ij}}{\rho}}$$

where V is the speed of sound, C_{ij} is the elastic constant or bulk modulus, and ρ is the material density. It is worth noting that the elastic constant differs depending on direction in anisotropic materials and the ij subscripts indicate that this directionality must be taken into account.

The speed of sound also varies with temperature since material characteristics are often dependent on temperature. At $0°C$, the speed of sound is 331m/s in air, 1402m/s in water and 6420m/s in Aluminum. Even though water is much denser than air, it is a lot more incompressible than air and its bulk modulus is larger than that of air by a larger factor. Therefore sound travels much faster in water than in air.

In the case of a stretched string, the velocity of a transverse wave passing through it is given by

$$v = \sqrt{\frac{\tau}{\mu}}$$

where τ is the tension in the string and μ is its linear density.

Skill 5.3   Analyzing reflection and transmission of mechanical waves

Mechanical waves can be described as a propagating disturbance in a medium or local oscillation of a material. Mechanical waves transmit energy, but the medium itself does not propagate. For example, a wave can travel on a slinky, but the slinky as a unit does not move. Since mechanical waves require a medium, they are unlike electromagnetic waves because they can not exist in a vacuum.

Mechanical waves may be transverse or longitudinal. Returning to the Slinky example, the individual atoms can be made to oscillate from their equilibrium positions in two distinct directions. In transverse waves, the displacement of the medium's particles is perpendicular to the direction of wave propagation. In longitudinal waves, the displacement of atoms or molecules is in the same direction as the wave's propagation. Sound waves are compression or longitudinal waves.

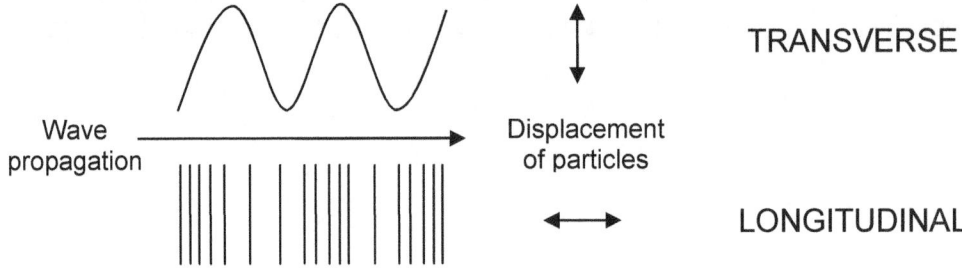

When a wave approaches the end of one medium, several boundary behaviors are possible. Depending on the change of medium, transmission and/or reflection may occur. To discuss boundary behavior, imagine a single pulse generated along a string. Four scenarios are important to consider. In each case, the incident pulse is the one approaching the boundary, while the reflected pulse is the wave resulting from interaction at the boundary.

First, consider a pulse approaching a fixed end or rigid point. In this case, the wave will be totally reflected and inverted. (Left)

Second, consider a pulse approaching a loose end or movable point. In this case, the wave will be totally reflected, but not inverted. (Right)

Pulses Approaching Different Boundaries

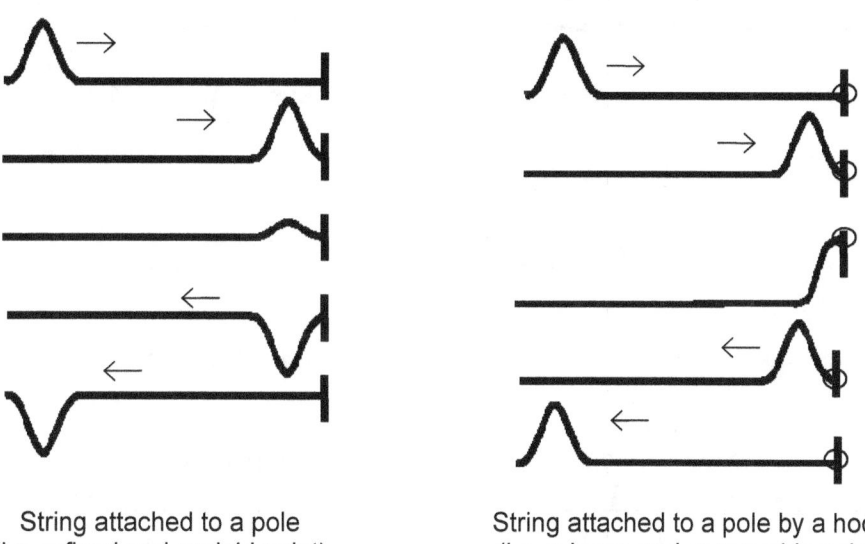

String attached to a pole
(i.e. a fixed end or rigid point)
Inverted!

String attached to a pole by a hoop
(i.e. a loose end or movable point)
NOT Inverted!

The third and fourth scenarios involve situation where the pulse on a string is crossing between medium of different densities.

Imagine a pulse traveling from a less dense medium towards a denser medium (i.e. from a thin rope connected to a thicker rope). In this case, the wave is partially transmitted and partially reflected. The transmitted pulse will have less velocity and smaller wavelength. The reflected pulse will have the same velocity and wavelength as the original incident pulse, but it will be inverted. (Left)

Imagine a pulse traveling from a less dense medium towards a denser medium (i.e. from a thick rope connected to a thinner rope). In this case, the wave is partially transmitted and partially reflected. The transmitted pulse will have greater velocity and larger wavelength. The reflected pulse will have the same velocity and wavelength as the incident pulse, and is not inverted. (Right)

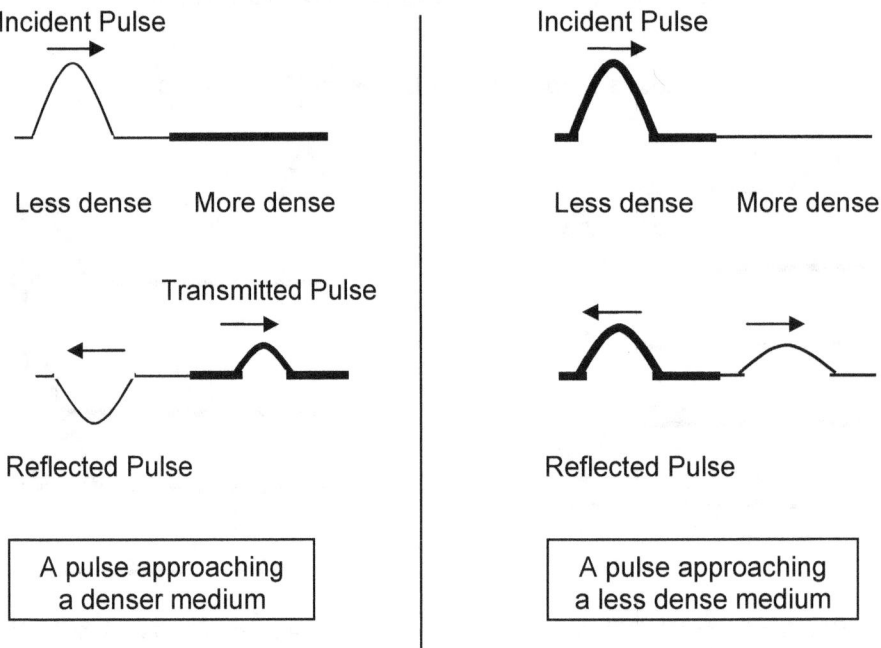

Boundary Behavior: The Partial Transmission and Reflection of Pulse Approaching Medium of Different Density

Careful consideration reveals that the fixed and loose end scenarios are the extrema conditions of the last two scenarios about differing medium density. For example, if a pulse is traveling from less dense medium into an extremely dense medium, then eventually the situation becomes analogous to a pulse approaching a fixed end or rigid point.

It is helpful to remember the following points when analyzing the transmission and reflection of mechanical waves:

1. Frequency is not changed when crossing a boundary
2. Wave speed is greater in the less dense medium
3. Wavelength is greater in the less dense medium
4. Pulse inversion occurs when the wave approaches a boundary where the new medium is more dense
5. The amplitude of the incident pulse is always greater than the amplitude of the reflected pulse.

Skill 5.4   Applying knowledge of the superposition principle to solve problems involving constructive and destructive interference

According to the **principle of linear superposition**, when two or more waves exist in the same place, the resultant wave is the sum of all the waves, i.e. the amplitude of the resulting wave at a point in space is the sum of the amplitudes of each of the component waves at that point. Interference is usually observed in coherent waves, well-correlated waves that have very similar frequencies or even come from the same source.

Superposition of waves may result in either constructive or destructive interference. **Constructive interference** occurs when the crests of the two waves meet at the same point in time. Conversely, **destructive interference** occurs when the crest of one wave and the trough of the other meet at the same point in time. It follows, then, that constructive interference increases amplitude and destructive interference decreased it. We can also consider interference in terms of wave phase; waves that are out of phase with one another will interfere destructively while waves that are in phase with one another will interfere constructively. In the case of two simple sine waves with identical amplitudes, for instance, amplitude will double if the waves are exactly in phase and drop to zero if the waves are exactly 180° out of phase.

Additionally, **interference can create a standing wave**, a wave in which certain points always have amplitude of zero. Thus, the wave remains in a constant position. Standing waves typically results when two waves of the same frequency traveling in opposite directions through a single medium are superposed. View an animation of how interference can create a standing wave at the following URL:

http://www.glenbrook.k12.il.us/GBSSCI/PHYS/mmedia/waves/swf.html

All wavelengths in the EM spectrum can experience interference but it is easy to comprehend instances of interference in the spectrum of visible light. One classic example of this is Thomas Young's double-slit experiment. In this experiment a beam of light is shone through a paper with two slits and a striated wave pattern results on the screen. The light and dark bands correspond to the areas in which the light from the two slits has constructively (bright band) and destructively (dark band) interfered.

Similarly, we may be familiar with examples of interference in sound waves. When two sounds waves with slightly different frequencies interfere with each other, beat results. We hear a beat as a periodic variation in volume with a rate that depends on the difference between the two frequencies. You may have observed this phenomenon when listening to two instruments being tuned to match; beating will be heard as the two instruments approach the same note and disappear when they are perfectly in tune.

Skill 5.5 Analyzing waves and solving problems involving amplitude, wavelength, period, frequency, and propagation speed in various media

All around us, there are examples of waves including ocean waves, sound, radio waves, microwaves, seismic waves, sunlight, x-rays, and radioactive gamma rays. Whether these waves actually displace media or simply carry energy, their positions fluctuate as they move through time and space. Often these fluctuations are regular and we can use both diagrams and mathematical equations to understand the pattern a wave follows in space and how quickly it moves.

Because mechanical, sound, and light waves surround us everyday, there are a variety of situations available for us to analyze. Though problems involving optics are more complicated we can solve simple problems involving waves by applying the definitions and equations from above. Below are several examples.

Problem:
A wave travels along a string. If it takes 0.30 seconds to move from its lowest to highest point, what is the frequency of the wave? If the wavelength is 1.8 m, what is the speed of the wave?

Solution:
Since 0.30 seconds are required for the wave to move from its lowest to highest point, the period must be 0.60 seconds. We can find frequency from period:

$$f = \frac{1}{T} = \frac{1}{0.6\,\text{sec}} = 1.67\,\text{sec}^{-1} = 1.67\,Hz$$

We already know the frequency and wavelength of the wave, so we can easily find the speed:

$$v = f \times \lambda = 1.67\,Hz \times 1.8\,m = 3.0\,\frac{m}{\sec}$$

Problem:
Waves are generated in two identical tanks of water. The amplitude of the wave in Tank 1 is 0.4 cm; in Tank 2 the amplitude is 0.8 cm. How much more energy is transported by the wave in Tank 2 than in Tank 1?

Solution:
Remember that the energy of a wave is proportional to the square of its amplitude. Thus, doubling the amplitude quadruples the energy, and the wave in Tank 2 transports four times as much energy as the one in Tank 1.

Problem:
A woman hiking in a canyon yells out "hello" and 2.4 seconds later hears her echo (the reflection of the sound wave off the canyon wall). If the speed of the sound wave is 345 m/sec, how far away is the canyon wall?

Solution:
If it takes 2.4 seconds for the echo to reach the woman, it must take half that time for the sound to reach canyon wall, that is, 1.2 seconds. Thus, we can find the distance to the wall:

$$d = v \times t = 345\,\frac{m}{\sec} \times 1.2\,\sec = 414\,m$$

Problem:
Two tuning forks are played simultaneously. If one fork has a frequency of 450 Hz and the other has a frequency of 446 Hz, how many beats will be heard after 10 seconds?

Solution
The beat frequency will be 450-446 Hz= 4 Hz

$$Number\ of\ beats = 4\,Hz \times 10\,\sec = 40\ beats$$

PHYSICS

Problem: The interference maxima (location of bright spots created by constructive interference) for double-slit interference are given by

$$\frac{n\lambda}{d} = \frac{x}{D} = \sin\theta \quad n=1,2,3...$$

where λ is the wavelength of the light, d is the distance between the two slits, D is the distance between the slits and the screen on which the pattern is observed and x is the location of the nth maximum. If the two slits are 0.1mm apart, the screen is 5m away from the slits, and the first maximum beyond the center one is 2.0 cm from the center of the screen, what is the wavelength of the light?

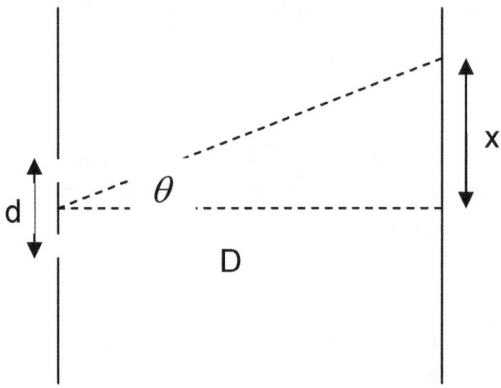

Solution: λ = xd/(Dn) = 0.02 x 0.0001/ (5 x1) = 400 nanometers

Potential energy=½ k A² cos² ($\omega_0 t+\Phi$)

Total energy=½ k A²

We know that sine and cosine are 90° out of phase. Thus, this is a mathematical statement that when potential energy is maximal, kinetic energy is minimum. Further, kinetic energy plus potential energy is always equal to a constant, that is, total energy in the system is always conserved.

COMPETENCY 6.0 UNDERSTAND THE FUNDAMENTAL PRINCIPLES OF LIGHT AND OPTICS

Skill 6.1 Analyzing properties of images produced by various mirrors

Plane mirrors form virtual images. In other words, the image is formed behind the mirror where light does not actually reach. The image size is equal to the object size and object distance is equal to the image distance; i.e. the image is the same distance behind the mirror as the object is in front of the mirror. Another characteristic of plane mirrors is left-right reversal.

Example: Suppose you are standing in front of a mirror with your right hand raised. The image in the mirror will be raising its left hand.

Problem: If a cat creeps toward a mirror at a rate of 0.20 m/s, at what speed will the cat and the cat's image approach each other?

Solution: In one second, the cat will be 0.20 meters closer to the mirror. At the same time, the cat's image will be 0.20 meters closer to the cat. Therefore, the cat and its image are approaching each other at the speed of 0.40 m/s.

Problem: If an object that is two feet tall is placed in front of a plane mirror, how tall will the image of the object be?

Solution: The image of the object will have the same dimensions as the actual object, in this case, a height of two feet. This is because the magnification of an image in a plane mirror is 1.

Curved mirrors

Curved mirrors are usually sections of spheres. In a concave mirror the inside of the spherical surface is silvered while in a convex mirror it is the outside of the spherical surface that is silvered.

Terminology associated with spherical mirrors:

Principal axis: The line joining the center of the sphere (of which we imagine the mirror is a section) to the center of the reflecting surface.

Center of curvature: The center of the sphere of which the mirror is a section.

Vertex: The point on the mirror where the principal axis meets the mirror or the geometric center of the mirror.

Focal point: The point at which light rays traveling parallel to the principal axis will meet after reflection in a concave mirror. For a convex mirror, it is the point from which light rays traveling parallel to the principal axis will appear to diverge from after reflection. The focal point is midway between the center of curvature and the vertex.

Focal length: The distance between the focal point and the vertex.

Radius of curvature: The distance between the center of curvature and the vertex, i.e. the radius of the sphere of which the mirror is a section. The radius of curvature is twice the focal length.

Image characteristics for **concave mirrors**:

1) If the object is located **beyond the center of curvature**, the image will be **real, inverted, smaller** and located between the focal point and center of curvature.
2) If the object is located **at the center of curvature**, the image will be **real, inverted, of the same height** and also located at the center of curvature.
3) If the object is located **between the center of curvature and focal point**, the image will be **real, inverted, larger** and located beyond the center of curvature.
4) If the object is located **at the focal point no image is formed**.
5) If the object is located **between the focal point and vertex**, the image will be **virtual, upright, larger** and located on the opposite side of the mirror.
6) If the object is located **at infinity** (very far away), the image is **real, inverted, smaller** and located at the focal point.

For **convex mirrors**, the image is always **virtual, upright, reduced in size** and formed on the opposite side of the mirror.

For image characteristics of curved mirrors see. The relationship between the object distance from vertex s, the image distance from vertex s', and the focal length f is given by the equation

$$\frac{1}{s} + \frac{1}{s'} = \frac{1}{f}$$

Magnification is defined as:

$$m = \frac{y'}{y} = -\frac{s'}{s}$$

where m=magnification, y'=image height, y=object height

Problem: A concave mirror collects light from a star. If the light rays converge at 50 cm, what is the radius of curvature of the mirror?

Solution: The point at which the rays converge is known as the focal point. The focal length, in this case, 50 cm, is the distance from the focal point to the mirror. The radius of curvature is the distance from the vertex to the center of curvature. The vertex is the point on the mirror where the principal axis meets the mirror. The center of curvature represents the point in the center of the sphere from which the mirror was sliced. Since the focal point is the midpoint of the line from the vertex to the center of curvature, or focal length, the focal length would be one-half the radius of curvature. Since the focal length in this case is 50 cm, the radius of curvature would be 100 cm.

Problem: An image of an object in a mirror is upright and reduced in size. In what type of mirror is this image being viewed, plane, concave, or convex?

Solution: The image in a plane mirror would be the same size as the object. The image in a concave mirror would be magnified if upright. Only a convex mirror would produce a reduced upright image of an object.

Skill 6.2 Applying knowledge of ray diagrams and Snell's law to solve problems involving refraction

Ray diagrams are a convenient way to visualize the propagation of waves and to perform reasonably accurate calculations of the effect of mirrors and lenses on light.

For mirrors, the focal point is either real (for concave mirrors) or virtual (for convex mirrors). In either case, the focal point is found by looking at the behavior of two parallel rays incident upon the mirror.

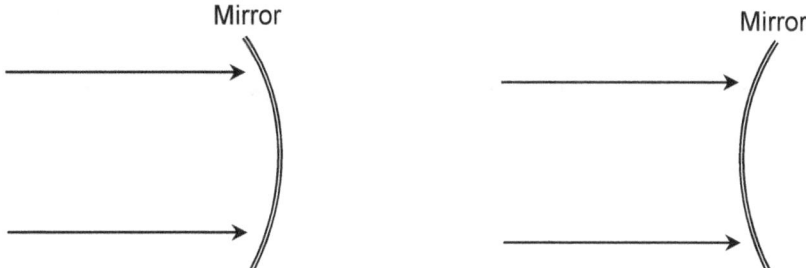

For each incident ray, the angle of reflection θ_r is equal to the angle of incidence θ_i, where the angle is measured from the normal to the mirror surface at the point of incidence.

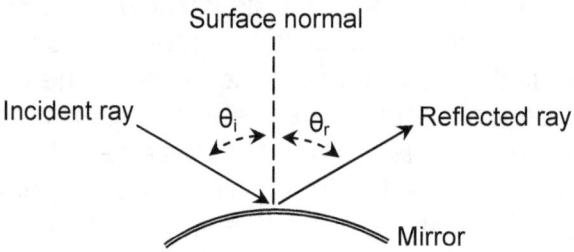

When this law is applied to the rays for either mirror, the focal points (f) are revealed as the (real or virtual) intersection of the rays. The virtual focus point is the intersection for the reflected rays when they are extended beyond the surface of the mirror.

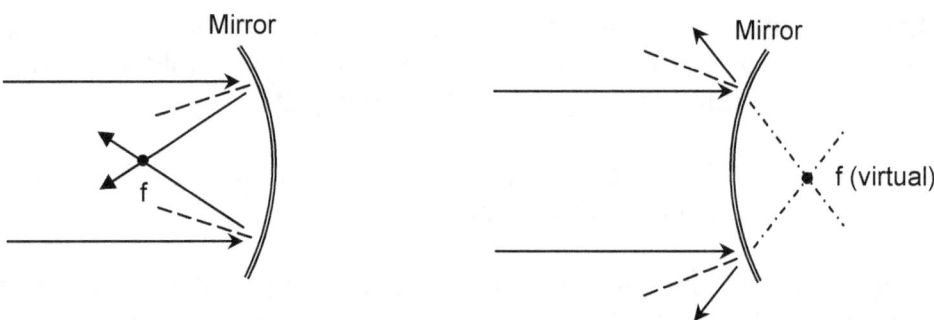

The focal point of a lens is found in a similar manner, with the exception that instead of using the law of reflection, refraction by way of Snell's law must be applied. Since real lenses have a finite thickness, Snell's law must be used for the ray both as it enters the lens material and as it exits the lens material.

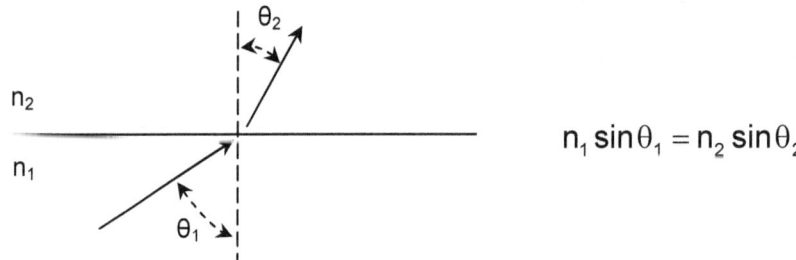

$$n_1 \sin\theta_1 = n_2 \sin\theta_2$$

As with reflection, refraction requires consideration of the normal to the surface. Also, the refractive indices must be used. It is assumed here that the refractive index of the outer medium is n_1 and the refractive index of the lens is n_2, and that $n_1 < n_2$. The intersection of parallel incident rays determines the focal point, which may again be either real or virtual. Real focal points occur on the opposite side of the lens from the source of illumination, and virtual focal points occur on the same side as the source of illumination.

Only one lens case is shown here, but the principles apply equally to all variations of lenses.

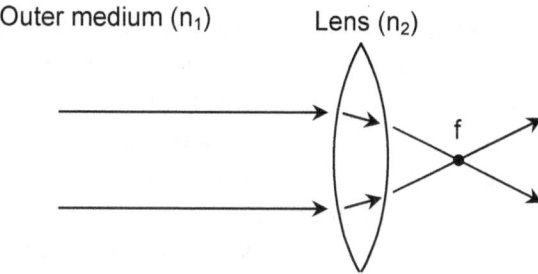

Care must be taken in properly identifying the normal and in applying Snell's law to both incidences of refraction for each ray.

Determining the point of image formation for mirrors and lenses is accomplished in a similar manner to that of the focal points. As with the focal points, image points may be either real or virtual, depending on the characteristics of the mirror or lens. To find the image point, two rays of differing angles must be traced as they interact with the lens or mirror. The initial directions of the rays can be chosen arbitrarily, but it is ideal to choose the directions such that the difficulty with determining the direction of the reflected or refracted ray is minimized. The intersection of these two rays is the image point. Only two examples are shown here, but the principles behind these examples may be applied to any variation of the situations, as well as to any combination of lenses and mirrors.

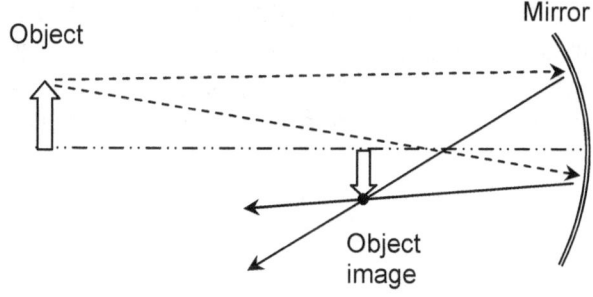

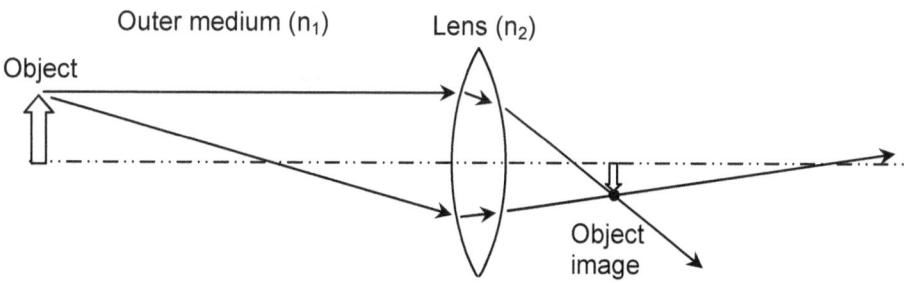

Skill 6.3 Analyzing properties or images produced by convex and concave lenses

A lens is a device that causes electromagnetic radiation to converge or diverge. The most familiar lenses are made of glass or plastic and designed to concentrate or disperse visible light. Two of the most important parameters for a lens are its thickness and its focal length. Focal length is a measure of how strongly light is concentrated or dispersed by a lens. For a convex or converging lens, the focal length is the distance at which a beam of light will be focused to a single spot. Conversely, for a concave or diverging lens, the focal length is the distance to the point from which a beam appears to be diverging.

The images produced by lenses can be either virtual or real. A virtual image is one that is created by rays of light that appear to diverge from a certain point. Virtual images cannot be seen on a screen because the light rays do not actually meet at the point where the image is located. If an image and object appear on the same side of a converging lens, that image is defined as virtual. For virtual images, the image location will be negative and the magnification positive. Real images, on the other hand, are formed by light rays actually passing through the image. Thus, real images are visible on a screen. Real images created by a converging lens are inverted and have a positive image location and negative magnification.

A thin lens in one in which focal length is much greater than lens thickness. For problems involving thin lenses, we can disregard any optical effects of the lens itself. Additionally, we can assume that the light that interacts with the lens makes a small angle with the optical axis of the system and so the sine and tangent values of the angle are approximately equal to the angle itself. This paraxial approximation, along with the thin lens assumptions, allows us to state:

$$\frac{1}{s} + \frac{1}{s'} = \frac{1}{f}$$

Where s=distance from the lens to the object (object location)
s'=distance from the lens to the image (image location)
f=focal length of the lens

Most lenses also cause some magnification of the object. Magnification is defined as:

$$m = \frac{y'}{y} = -\frac{s'}{s}$$

Where m=magnification
y'=image height
y=object height

Sign conventions will make it easier to understand thin lens problems:

Focal length: positive for a converging lens; negative for a diverging lens
Object location: positive when in front of the lens; negative when behind the lens
Image location: positive when behind the lens; negative when in front of the lens
Image height: positive when upright; negative when upside-down.
Magnification: positive for an erect, virtual image; negative for an inverted, real image

Problem:
A converging lens has a focal length of 10.00 cm and forms a 2.0 cm tall image of a 4.00 mm tall real object to the left of the lens. If the image is erect, is the image real or virtual? What are the locations of the object and the image?

Solution:
We begin by determining magnification:

$$m = \frac{y'}{y} = \frac{0.02m}{0.004m} = 5$$

Since the magnification is positive and the image is erect, we know the image must be virtual.

To find the locations of the object and image, we first relate them by using the magnification:

$$m = -\frac{s'}{s}$$
$$s' = -ms$$

Then we substitute into the thin lens equation, creating one variable in one unknown:

$$\frac{1}{s} + \frac{1}{s'} = \frac{1}{f}$$

$$\frac{1}{s} - \frac{1}{5s} = \frac{1}{10cm}$$

$$\frac{5-1}{5s} = \frac{1}{10cm}$$

$$s = \frac{40cm}{5} = 8cm \rightarrow \rightarrow s' = -5 \times 8cm = -40cm$$

Thus the object is located 8 cm to the left of the lens and the image is 40 cm to the left of the lens.

TEACHER CERTIFICATION STUDY GUIDE

Skill 6.4   Analyzing the phenomena of dispersion, diffraction, and polarization

Wave **refraction** is a change in direction of a wave due to a change in its speed. This most commonly occurs when a wave passes from one material to another, such as a light ray passing from air into water or glass. However, light is only one example of refraction; any type of wave can undergo refraction. Another example would be physical waves passing from water into oil. At the boundary of the two media, the wave velocity is altered, the direction changes, and the wavelength increases or decreases. However, the frequency remains constant.

The **index of refraction**, n, is the amount by which light slows in a given material and is defined by the formula

$$n = \frac{c}{v}$$

where v represents the speed of light through the given material.

Problem: The speed of light in an unknown medium is measured to be $1.24 \times 10^8 \, m/s$. What is the index of refraction of the medium?

Solution:

$$n = \frac{c}{v}$$

$$n = \frac{3.00 \times 10^8}{1.24 \times 10^8} = 2.42$$

Referring to a standard table showing indices of refraction, we would see that this index corresponds to the index of refraction for diamond.

Snell's Law describes how light bends, or refracts, when traveling from one medium to the next. It is expressed as

$$n_1 \sin \theta_1 = n_2 \sin \theta_2$$

where n_i represents the index of refraction in medium i, and θ_i represents the angle the light makes with the normal in medium i.

Reflection is the change in direction of a wave at an interface between two dissimilar media such that the wave returns into the medium from which it originated. The most common example of this is light waves reflecting from a mirror, but sound and water waves can also be reflected. **The law of reflection states that the angle of incidence is equal to the angle of reflection.**

PHYSICS

Problem: The index of refraction for light traveling from air into an optical fiber is 1.44. (a) In which direction does the light bend? (b) What is the angle of refraction inside the fiber, if the angle of incidence on the end of the fiber is 22°?

Solution: (a) The light will bend toward the normal since it is traveling from a rarer region (lower n) to a denser region (higher n).

(b) Let air be medium 1 and the optical fiber be medium 2:

$$n_1 \sin\theta_1 = n_2 \sin\theta_2$$
$$(1.00)\sin 22° = (1.44)\sin\theta_2$$
$$\sin\theta_2 = \frac{1.00}{1.44}\sin 22° = (.6944)(.3746) = 0.260$$
$$\theta_2 = \sin^{-1}(0.260) = 15°$$

The angle of refraction inside the fiber is $15°$.

Reflection may occur whenever a wave travels from a medium of a given refractive index to another medium with a different index. A certain fraction of the light is reflected from the interface and the remainder is refracted. However, when the wave is moving from a dense medium into one less dense, that is the refractive index of the first is greater than the second, a **critical angle** exists which will create a phenomenon known as **total internal reflection.** In this situation all of the wave is reflected. The critical angle of incidence θ_c is the one for which the angle of reflection is 90 degrees. Thus, according to Snell's law

$$n_1 \sin\theta_c = n_2 \sin 90° \Rightarrow \theta_c = \sin^{-1}\frac{n_2}{n_1}$$

Snell's law may also be used to understand the phenomenon of dispersion since it relates the angle of refraction at the boundary between two media to the relative refractive indices. Since different frequency components of visible light have different indices of refraction in any medium other than vacuum, each component of a beam of white light is refracted at a different angle when it crosses a surface resulting in a separation of the colors.

Diffraction occurs when part of a wave front is obstructed. Diffraction and interference are essentially the same physical process. Diffraction refers to various phenomena associated with wave propagation such as the bending, spreading, and interference of waves emerging from an aperture. It occurs with any type of wave including sound waves, water waves, and electromagnetic waves such as light and radio waves.

Dispersion is the separation of a wave into its constituent wavelengths due to interaction with a material occurring in a wavelength-dependent manner (as in thin-film interference for instance). Dispersive prisms separate white light into these constituent colors by relying on the differences in refractive index that result from the varying frequencies of the light. Prisms rely on the fact that light changes speed as it moves from one medium to another. This then causes the light to be bent and/or reflected. The degree to which bending/reflection occurs is a function of the light's angle of incidence and the refractive indices of the media.

Polarization is a property of transverse waves that describes the plane perpendicular to the direction of travel in which the oscillation occurs. In unpolarized light, the transverse oscillation occurs in all planes perpendicular to the direction of travel. Polarized light (created, for instance, by using polarizing filters that absorb light oscillating in other planes) oscillates in only a selected plane. An everyday example of polarization is found in polarized sunglasses which reduce glare.

Skill 6.5   Analyzing wave properties of the electromagnetic spectrum

The electromagnetic spectrum is measured using frequency (f) in hertz or wavelength (λ) in meters. The frequency times the wavelength of every electromagnetic wave equals the speed of light (3.0×10^8 meters/second).

Roughly, the range of wavelengths of the electromagnetic spectrum is:

	f	λ
Radio waves	$10^5 - 10^{-1}$ hertz	$10^3 - 10^9$ meters
Microwaves	$3\times10^9 - 3\times10^{11}$ hertz	$10^{-3} - 10^{-1}$ meters
Infrared radiation	$3\times10^{11} - 4\times10^{14}$ hertz	$7\times10^{-7} - 10^{-3}$ meters
Visible light	$4\times10^{14} - 7.5\times10^{14}$ hertz	$4\times10^{-7} - 7\times10^{-7}$ meters
Ultraviolet radiation	$7.5\times10^{14} - 3\times10^{16}$ hertz	$10^{-8} - 4\times10^{-7}$ meters
X-Rays	$3\times10^{16} - 3\times10^{19}$ hertz	$10^{-11} - 10^{-8}$ meters
Gamma Rays	$>3\times10^{19}$ hertz	$<10^{-11}$ meters

Radio waves are used for transmitting data. Common examples are television, cell phones, and wireless computer networks. Microwaves are used to heat food and deliver Wi-Fi service. Infrared waves are utilized in night vision goggles. Visible light we are all familiar with as the human eye is most sensitive to this wavelength range. UV light causes sunburns and would be even more harmful if most of it were not captured in the Earth's ozone layer. X-rays aid us in the medical field and gamma rays are most useful in the field of astronomy.

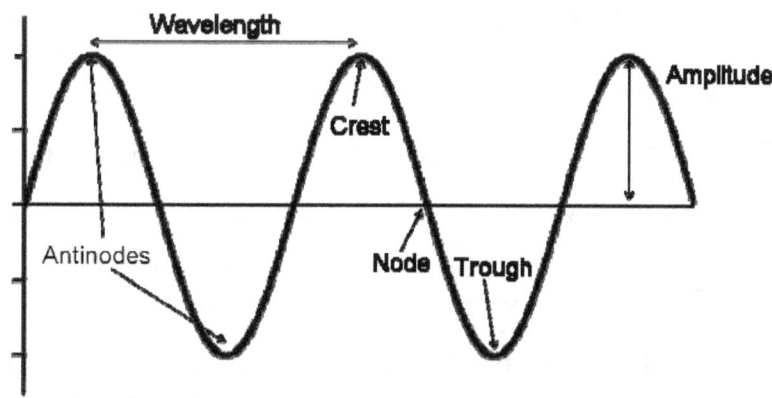

This diagram of a sinusoidal wave shows us displacement caused by the wave as it propagates through a medium. This displacement can be graphed against either time or distance. Note how displacement depends on the distance which the wave has traveled/ how much time has elapsed. So if we chose a particular displacement (let's say the crest), the wave will return to that displacement value (i.e., crest again) after one period (T) or one wavelength (λ).

The general equation for a wave is a partial differential equation which can be simplified to express the behavior of commonly encountered harmonic waveforms. For instance, for a standing wave:

$$y(z,t) = A(z,t)\sin(kz - \omega t + \phi)$$

    where y=displacement
        z=distance
        t=time
        k=wave number
        ω=angular frequency
        ϕ=phase
        A(z,t)=the amplitude envelope of the wave

The important concept to note is that y is a function of both z and t. This means that the wave's position depends on both time and distance, just as was seen in the diagram above.

To fully understand waves, it is important to understand many of the terms used to characterize them.

Wave velocity: Two velocities are used to describe waves. The first is phase velocity, which is the rate at which a wave propagates. For instance, if you followed a single crest of a wave, it would appear to move at the phase velocity. The second type of velocity is known as group velocity and is the speed at which variations in the wave's amplitude shape propagate through space. Group velocity is often conceptualized as the velocity at which energy is transmitted by a wave. Phase velocity is denoted v_p and group velocity is denoted v_g.

Crest: The maximum value that a wave assumes; the highest point.

Trough: The lowest value that a wave assumes; the lowest point.

Nodes: The points on a wave with minimal amplitude.

Antinodes: The farthest point from the node on the amplitude axis; both the crests and the troughs are antinodes.

Amplitude: The distance from the wave's highest point (the crest) to the equilibrium point. This is a measure of the maximum disturbance caused by the wave and is typically denoted by A.

Wavelength: The distance between any two sequential troughs or crests denoted λ and representing a complete cycle in the repeated wave pattern.

Period: The time required for a complete wavelength or cycle to pass a given point. The period of a wave is usually denoted T.

Frequency: The number of periods or cycles per unit time (usually a second). The frequency is denoted f and is the inverse of the wave's period (that is, $f=1/T$).

Phase: This is a given position in the cycle of the wave. It is most commonly used in discussing a "being out of phase" or a "phase shift", an offset between waves.

Longitudinal waves: These are created by oscillations in the direction in which the wave travels. Thus, if we imagine a longitudinal waveform moving down a tube, particles will move back and forth parallel to the sides of the tube.

Transverse waves: These waves, on the other hand, oscillate in a direction perpendicular to the direction of wave travel. So let's imagine the same tube, this time with a transverse wave traveling down it. In this case, the particles oscillate up and down or side to side within the tube. Particle displacement in a transverse wave can also be easily visualized in the vibration of a taut string.

Polarization: A property of transverse waves that describes the plane perpendicular to the direction of travel in which the oscillation occurs. Note that longitudinal waves are not polarized because they can oscillate only in one direction, the direction of travel.

TEACHER CERTIFICATION STUDY GUIDE

COMPETENCY 7.0 UNDERSTAND THE PRINCIPLES OF THERMODYNAMICS

Skill 7.1 Differentiating between thermal energy and temperature and solving problems involving thermal energy

Heat is generally measured in terms of temperature, a measure of the average internal energy of a material. Temperature is an **intensive property**, meaning that it does not depend on the amount of material. Heat content is an **extensive property** because more material at the same temperature will contain more heat. The relationship between the change in heat energy of a material and the change in its temperature is given by $\Delta Q = mC\Delta T$, where ΔQ is the change in heat energy, m is the mass of the material, ΔT is the change in temperature and C is the specific heat which is characteristic of a particular material.

The **zeroth law of thermodynamics** generally deals with bodies in thermal equilibrium with each other and is the basis for the idea of temperature. Most commonly, the law is stated as, "If two thermodynamic systems are in thermal equilibrium with a third, they are also in thermal equilibrium with each other." Said another way, this very basic law simply states that if object A is same temperature as object B, and object C is the same temperature as object B, then object A and C are also the same temperature.

Skill 7.2 Identifying methods of thermal energy transfer

All heat transfer is the movement of thermal energy from hot to cold matter. This movement down a thermal gradient is a consequence of the second law of thermodynamics. The three methods of heat transfer are listed and explained below.

Conduction: Electron diffusion or photo vibration is responsible for this mode of heat transfer. The bodies of matter themselves do not move; the heat is transferred because adjacent atoms that vibrate against each other or as electrons flow between atoms. This type of heat transfer is most common when two solids come in direct contact with each other. This is because molecules in a solid are in close contact with one another and so the electrons can flow freely. It stands to reason, then, that metals are good conductors of thermal energy. This is because their metallic bonds allow the freest movement of electrons. Similarly, conduction is better in denser solids. Examples of conduction can be seen in the use of copper to quickly convey heat in cooking pots, the flow of heat from a hot water bottle to a person's body, or the cooling of a warm drink with ice.

PHYSICS

Convection: Convection involves some conduction but is distinct in that it involves the movement of warm particles to cooler areas. Convection may be either natural or forced, depending on how the current of warm particles develops. Natural convection occurs when molecules near a heat source absorb thermal energy (typically via conduction), become less dense, and rise. Cooler molecules then take their place and a natural current is formed. Forced convection, as the name suggests, occurs when liquids or gases are moved by pumps, fans, or other means to be brought into contact with warmer or cooler masses. Because the free motion of particles with different thermal energy is key to this mode of heat transfer, convection is most common in liquid and gases. Convection can, however, transfer heat between a liquid or gas and a solid. Forced convection is used in "forced air" home heating systems and is common in industrial manufacturing processes. Additionally, natural convection is responsible for ocean currents and many atmospheric events. Finally, natural convection often arises in association with conduction, for instance in the air near a radiator or the water in a pot on the stove.

Radiation: This method of heat transfer occurs via electromagnetic radiation. All matter warmer than absolute zero (that is, all known matter) radiates heat. This radiation occurs regardless of the presence of any medium. Thus, it occurs even in a vacuum. Since light and radiant heat are both part of the EM spectrum, we can easily visualize how heat is transferred via radiation. For instance, just like light, radiant heat is reflected by shiny materials and absorbed by dark materials. Common examples of radiant heat include the way sunlight travels from the sun to warm the earth, the use of radiators in homes, and the warmth of incandescent light bulbs.

Skill 7.3 Applying knowledge of thermodynamic work and the law of conservation of energy to solve a variety of problems

The first law of thermodynamics is a restatement of conservation of energy, i.e. the principle that energy cannot be created or destroyed. It also governs the behavior of a system and its surroundings. The change in heat energy supplied to a system (Q) is equal to the sum of the change in the internal energy (U) and the change in the work (W) done by the system against internal forces.

The internal energy of a material is the sum of the total kinetic energy of its molecules and the potential energy of interactions between those molecules. Total kinetic energy includes the contributions from translational motion and other components of motion such as rotation. The potential energy includes energy stored in the form of resisting intermolecular attractions between molecules. Mathematically, we can express the relationship between the heat supplied to a system, its internal energy and work done by it as

$$\Delta Q = \Delta U + \Delta W$$

Let us examine a sample problem that relies upon this law.

Problem: A closed tank has a volume of 40.0 m³ and is filled with air at 25°C and 100 kPa. We desire to maintain the temperature in the tank constant at 25°C as water is pumped into it. How much heat will have to be removed from the air in the tank to fill the tank ½ full?

Solution: The problem involves isothermal compression of a gas, so $\Delta U_{gas}=0$. Consulting the equation above, $\Delta Q = \Delta U + \Delta W$, it is clear that the heat removed from the gas must be equal to the work done by the gas.

$$Q_{gas} = W_{gas} = P_{gas}V_1 \ln\left(\frac{V_2}{V_T}\right) = P_{gas}V_T \ln\left(\frac{\frac{1}{2}V_T}{V_T}\right) = P_{gas}V_T \ln \frac{1}{2}$$

$$= (100 kPa)(40.0 m^3)(-0.69314) = -2772.58 kJ$$

Thus, the gas in the tank must lose 2772.58 kJ to maintain its temperature.

Skill 7.4 Demonstrating knowledge of the second law of thermodynamics

To understand the **second law of thermodynamics**, we must first understand the concept of entropy. Entropy is the transformation of energy to a more disordered state and is the measure of how much energy or heat is available for work. The greater the entropy of a system, the less energy is available for work. The simplest statement of the second law of thermodynamics is that the entropy of an isolated system not in equilibrium tends to increase over time. The entropy approaches a maximum value at equilibrium. Below are several common examples in which we see the manifestation of the second law.

- The diffusion of molecules of perfume out of an open bottle
- Even the most carefully designed engine releases some heat and cannot convert all the chemical energy in the fuel into mechanical energy
- A block sliding on a rough surface slows down
- An ice cube sitting on a hot sidewalk melts into a little puddle; we must provide energy to a freezer to facilitate the creation of ice

When discussing the second law, scientists often refer to the "arrow of time". This is to help us conceptualize how the second law forces events to proceed in a certain direction. To understand the direction of the arrow of time, consider some of the examples above; we would never think of them as proceeding in reverse. That is, as time progresses, we would never see a puddle in the hot sun spontaneously freeze into an ice cube or the molecules of perfume dispersed in a room spontaneously re-concentrate themselves in the bottle. The above-mentioned examples are **spontaneous** as well as **irreversible**, both characteristic of increased entropy. Entropy change is zero for a complete cycle in a **reversible process**, a process where infinitesimal quasi-static changes in the absence of dissipative forces can bring a system back to its original state without a net change to the system or its surroundings. All real processes are irreversible. The idea of a reversible process, however, is a useful abstraction that can be a good approximation in some cases.

A quantitative measure of entropy S is given by the statement that the change in entropy of a system that goes from one state to another in an isothermal and reversible process is the amount of heat absorbed in the process divided by the absolute temperature at which the process occurs.

$$\Delta S = \frac{\Delta Q}{T}$$

Stated more generally, the entropy change that occurs in a state change between two equilibrium states A and B via a reversible process is given by

$$\Delta S_{A \to B} = \int_A^B \frac{dQ}{T}$$

Problem: What is the change in entropy of a cube of ice of mass 30g which melts at temperature 0C? The latent heat of fusion of ice is 334 KJ/Kg.

Solution: The amount of heat absorbed by the ice cube $30 \times 10^{-3} \times 334$ KJ = 10020 J. Thus change in entropy = (10020/273)J/K = 36.7 J/K

The second law of thermodynamics may also be stated in the following ways:
 1. No machine is 100% efficient.
 2. Heat cannot spontaneously pass from a colder to a hotter object.

If we consider a **heat engine** that absorbs heat Q_h from a hot reservoir at temperature T_h and does work W while rejecting heat Q_c to a cold reservoir at a lower temperature T_c, $Q_h - Q_c$ = W. The efficiency of the engine is the ratio of the work done to the heat absorbed and is given by

$$\varepsilon = \frac{W}{Q_h} = \frac{Q_h - Q_c}{Q_h} = 1 - \frac{Q_c}{Q_h}$$

It is impossible to build a heat engine with 100% efficiency (i.e. one where $Q_c=0$).

Carnot described an ideal reversible engine, the **Carnot engine**, that works between two heat reservoirs in a cycle known as the **Carnot cycle** which consists of two isothermal (12 and 34) and two adiabatic processes (23 and 41) as shown in the diagram below.

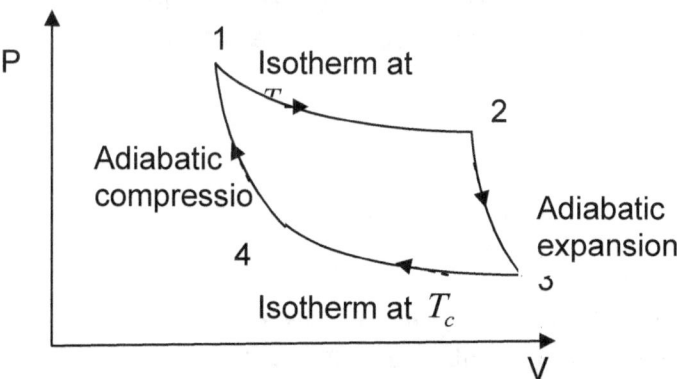

The efficiency of a Carnot engine is given by $\varepsilon = 1 - \dfrac{T_c}{T_h}$ where the temperature values are absolute temperatures. This is the highest efficiency that any engine working between T_c and T_h can reach.

According to **Carnot's theorem**, no engine working between two heat reservoirs can be more efficient than a reversible engine. All such reversible engines have the same efficiency.

Skill 7.5 Analyzing the states of matter and energy transformation during phase changes

A substance's molar **heat capacity** is the heat required to **change the temperature of one mole of the substance by one degree**. Heat capacity has units of joules per mol- kelvin or joules per mol- °C. The two units are interchangeable because we are only concerned with differences between one temperature and another. A Kelvin degree and a Celsius degree are the same size.

The **specific heat** of a substance (also called specific heat capacity) is the heat required to **change the temperature of one gram or kilogram by one degree.** Specific heat has units of joules per gram-°C or joules per kilogram-°C.

The temperature of a material rises when heat is transferred to it and falls when heat is removed from it. When the material is undergoing a phase change (e.g. from solid to liquid), however, it absorbs or releases heat without a corresponding change in temperature. The heat that is absorbed or released during phase change is known as latent heat.

A **temperature vs. heat graph** can demonstrate these relationships visually. One can also calculate the specific heat or latent heat of phase change for the material by studying the details of the graph.

Example: The plot below shows heat applied to 1g of ice at -40C. The horizontal parts of the graph show the phase changes where the material absorbs heat but stays at the same temperature. The graph shows that ice melts into water at 0C and the water undergoes a further phase change into steam at 100C.

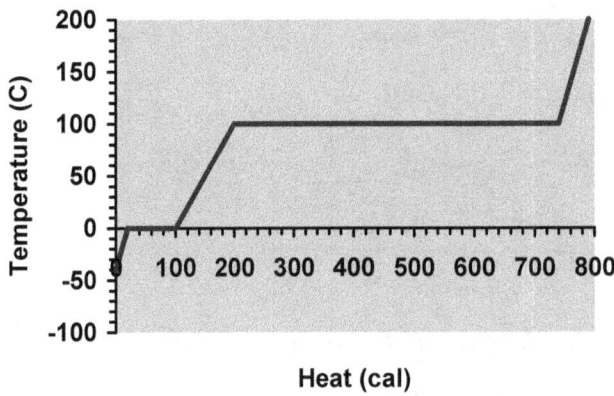

Heat (cal)

The specific heat of ice, water and steam and the latent heat of fusion and vaporization may be calculated from each of the five segments of the graph.

For instance, we see from the flat segment at temperature 0C that the ice absorbs 80 cal of heat. The latent heat L of a material is defined by the equation $\Delta Q = mL$ where ΔQ is the quantity of heat transferred and m is the mass of the material. Since the mass of the material in this example is 1g, the latent heat of fusion of ice is given by $L = \Delta Q / m$ = 80 cal/g.

The next segment shows a rise in the temperature of water and may be used to calculate the specific heat C of water defined by $\Delta Q = mC\Delta T$, where ΔQ is the quantity of heat absorbed, m is the mass of the material and ΔT is the change in temperature. According to the graph, ΔQ = 200-100 =100 cal and ΔT = 100-0=100C. Thus, C = 100/100 = 1 cal/gC.

Problem: The plot below shows the change in temperature when heat is transferred to 0.5g of a material. Find the initial specific heat of the material and the latent heat of phase change.

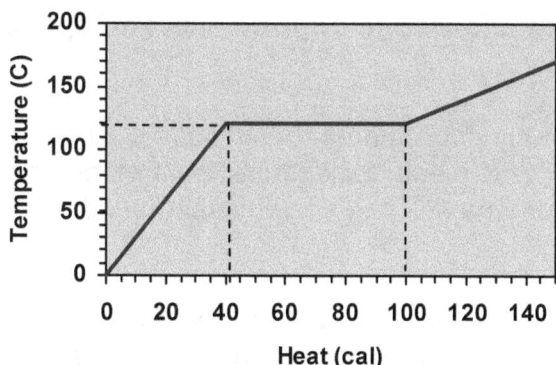

Solution:
Looking at the first segment of the graph, we see that ΔQ = 40 cal and ΔT = 120 C. Since the mass m = 0.5g, the specific heat of the material is given by
$C = \Delta Q / (m \Delta T)$ = 40/(0.5 X 120) = 0.67 cal/gC.

The flat segment of the graph represents the phase change. Here ΔQ = 100 - 40=60 cal. Thus, the latent heat of phase change is given by
$L = \Delta Q / m$ = 60/(0.5) = 120 cal/g.

SUBAREA III. ELECTRICITY, MAGNETISM, AND ATOMIC PHYSICS

COMPETENCY 8.0 UNDERSTAND ELECTRIC CHARGE AND INTERACTIONS BETWEEN CHARGED OBJECTS

Skill 8.1 Demonstrating knowledge of principles and application of electrostatics

Electrostatics involves the study of problems that include only stationary electric charges. Although most matter is electrically neutral on a macroscopic scale, the effects of electrostatic phenomena still can be witnessed in common, everyday experiences.

Perhaps the most common example is so-called **static electricity** which results from a buildup of charge on two separate objects or different regions of the same object. The excess charge, and corresponding electric potential, will repel objects with an excess charge of the same type (i.e., both positive or both negative) and attract objects of dissimilar type (i.e., positive and negative). For example, a glass rod rubbed with a piece of silk acquires a net positive charge due to a loss of some of (negatively charged) electrons. A plastic rod rubbed with wool acquires a net negative charge due to a deposit of excess electrons from the wool. These two rods, since they are of opposite charge, will attract one another.

The ability to transfer electrons is based on an object's position on the **triboelectric series**. If an object gives up electrons, it is on the positive end of the triboelectric series; if an object acquires electrons it is on the negative end of the triboelectric series. As one looks at the triboelectric series, one can determine which object will acquire electrons and which object will lose electrons if rubbed together. The further away two objects are from each other on the series, the greater the net charge will be between them, therefore, the greater the attraction will be between them. Below is a triboelectric series of common items.

A Triboelectric Series
Positive end of series
Dry human skin
Leather
Rabbit fur
Glass
Human Hair
Nylon
Wool
Lead
Cat fur
Silk
Aluminum
Paper
Cotton - Neutral
Steel - Neutral
Wood
Amber
Rubber
Nickel/copper
Brass/silver
Gold/Platinum
Polyester
Styrofoam
Saran Wrap
Scotch Tape
Vinyl
Silicon
Teflon
Negative end of the series

Polarization of an object can take place when a charged object is brought in close proximity of an uncharged object. When this occurs, the charges within the object itself separate; the like charge moves away from the charged object and the opposite charge moves towards the charged object.

Applications of Electrostatics

There are examples of electrostatic phenomena all around us. In nature, for instance, lightning is one of the most dramatic. However, electrostatic principles are used in industry quite often. Electrostatic filters have been created to remove particles from exhaust. Xerox machines use electrostatics to create a copy of an image. Electrostatics is also used in food and ore purification as well as to separate objects composed of different materials.

Skill 8.2 Solving problems using Coulomb's law

The easiest way to begin analyzing electrostatic phenomenon and calculating this force is by considering two point charges. Let us say that the charge on the first point is Q_1, the charge on the second point is Q_2, and the distance between them is r. Their interaction is governed by Coulomb's Law which gives the formula for the force F as:

$$F = k\frac{Q_1 Q_2}{r^2}$$

where $k = 9.0 \times 10^9 \, \frac{N \cdot m^2}{C^2}$ (known as Coulomb's constant)

The charge is a scalar quantity, however, the force has direction. For two point charges, the direction of the force is along a line joining the two charges. Note that the force will be repulsive if the two charges are both positive or both negative and attractive if one charge is positive and the other negative. Thus, a negative force indicates an attractive force.

Problem: Three point charges are located at the vertices of a right triangle as shown below. Charges, angles, and distances are provided (drawing not to scale). Find the force exerted on the point charge A.

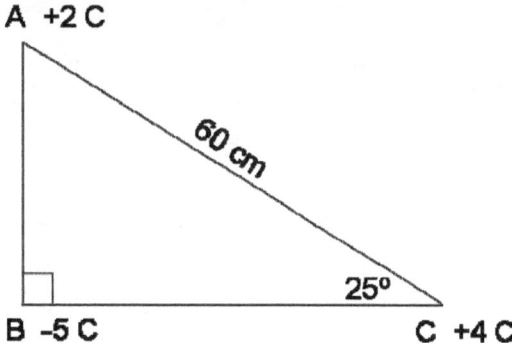

Solution: First we find the individual forces exerted on A by point B and point C. We have the information we need to find the magnitude of the force exerted on A by C.

$$F_{AC} = k\frac{Q_1 Q_2}{r^2} = 9\times 10^9 \frac{N\cdot m^2}{C^2}\left(\frac{4C \times 2C}{(0.6m)^2}\right) = 2\times 10^{11}\,N$$

To determine the magnitude of the force exerted on A by B, we must first determine the distance between them.

$$\sin 25° = \frac{r_{AB}}{60cm}$$
$$r_{AB} = 60cm \times \sin 25° = 25cm$$

Now we can determine the force.

$$F_{AB} = k\frac{Q_1 Q_2}{r^2} = 9\times 10^9 \frac{N\cdot m^2}{C^2}\left(\frac{-5C \times 2C}{(0.25m)^2}\right) = -1.4\times 10^{12}\,N$$

We can see that there is an attraction in the direction of B (negative force) and repulsion in the direction of C (positive force). To find the net force, we must consider the direction of these forces (along the line connecting any two point charges). We add them together using the law of cosines.

$$F_A^2 = F_{AB}^2 + F_{AC}^2 - 2F_{AB}F_{AC}\cos 75°$$
$$F_A^2 = (-1.4\times 10^{12}\,N)^2 + (2\times 10^{11}\,N)^2 - 2(-1.4\times 10^{12}\,N)(2\times 10^{11}\,N)^2 \cos 75°$$
$$F_A = 1.5\times 10^{12}\,N$$

This gives us the magnitude of the net force, now we will find its direction using the law of sines.

$$\frac{\sin\theta}{F_{AC}} = \frac{\sin 75°}{F_A}$$

$$\sin\theta = F_{AC}\frac{\sin 75°}{F_A} = 2\times 10^{11}\,N\frac{\sin 75°}{1.5\times 10^{12}\,N}$$

$$\theta = 7.3°$$

Thus, the net force on A is 7.3° west of south and has magnitude 1.5×10^{12} N. Looking back at our diagram, this makes sense, because A should be attracted to B (pulled straight south) but the repulsion away from C "pushes" this force in a westward direction.

Skill 8.3 Demonstrating knowledge of the electric field in the vicinity of point charges

Electric fields can be generated by a single point charge or by a collection of charges in close proximity. The electric field generated from a point charge is given by:

$$E = \frac{kQ}{r^2}$$

where E = the electric field

$k = 9.0 \times 10^9 \frac{N \cdot m^2}{C^2}$ (Coulomb's constant)

Q = the point charge
r = distance from the charge

Electric fields are visualized with field lines, which demonstrate the strength and direction of an electric field. The electric field around a positive charge points away from the charge and the electric field around a negative charge points toward the charge.

While it's easy enough to calculate and visualize the field generated by a single point charge, we can also determine the nature of an electric field produced by a collection of charge simply by adding the vectors from the individual charges. This is known as the superposition principle. The following equation demonstrates how this principle can be used to determine the field resulting from hundreds or thousands of charges.

$$\vec{E}_{total} = \sum_i \vec{E}_i = \vec{E}_1 + \vec{E}_2 + \vec{E}_3 \ldots$$

Skill 8.4 Solving basic problems involving electrostatic potential and electrostatic potential energy

Electric potential is simply the **potential energy** per unit of charge. Given this definition, it is clear that electric potential must be measured in joules per coulomb and this unit is known as a volt (J/C=V).

Within an electric field there are typically differences in potential energy. This **potential difference** may be referred to as **voltage**. The difference in electrical potential between two points is the amount of work needed to move a unit charge from the first point to the second point.

Stated mathematically, this is:

$$V = \frac{W}{Q}$$

where V= the potential difference
W= the work done to move the charge
Q= the charge

We know from mechanics, however, that work is simply force applied over a certain distance. We can combine this with Coulomb's law to find the work done between two charges distance r apart.

$$W = F \cdot r = k\frac{Q_1 Q_2}{r^2} \cdot r = k\frac{Q_1 Q_2}{r}$$

Now we can simply substitute this back into the equation above for electric potential:

$$V_2 = \frac{W}{Q_2} = \frac{k\frac{Q_1 Q_2}{r}}{Q_2} = k\frac{Q_1}{r}$$

Let's examine a sample problem involving electrical potential.

Problem: What is the electric potential at point A due to the 2 shown charges? If a charge of +2.0 C were infinitely far away, how much work would be required to bring it to point A?

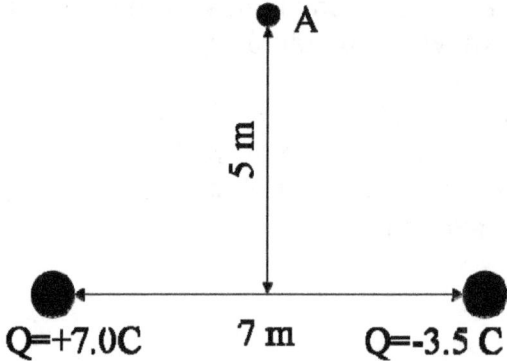

Solution: To determine the electric potential at point A, we simple find and add the potential from the two charges (this is the principle of superposition). From the diagram, we can assume that A is equidistant from each charge. Using the Pythagorean theorem, we determine this distance to be 6.1 m.

$$V = \frac{kq}{r} = k\left(\frac{7.0C}{6.1m} + \frac{-3.5C}{6.1m}\right) = 9 \times 10^9 \frac{N.m^2}{C^2}\left(0.57\frac{C}{m}\right) = 5.13 \times 10^9 V$$

Now, let's consider bringing the charged particle to point A. We assume that electric potential of these particle is initially zero because it is infinitely far away. Since now know the potential at point A, we can calculate the work necessary to bring the particle from V=0, i.e. the potential energy of the charge in the electrical field:

$$W = VQ = (5.13 \times 10^9) \times 2J = 10.26 \times 10^9 J$$

The large results for potential and work make it apparent how large the unit coulomb is. For this reason, most problems deal in microcoulombs (μC).

COMPETENCY 9.0 UNDERSTAND MAGNETS, MAGNETIC FIELDS, AND ELECTROMAGNETIC INDUCTION

Skill 9.1 Demonstrating knowledge of the properties of permanent magnets

A ferromagnetic material can be either fully magnetized (saturation) or completely unmagnetized. Ferromagnetic domain theory presents an explanation of this phenomenon by viewing the material as being divided into a number of small or microscopic "domains" each of which has its own associated magnetic dipole moment that is aligned in a particular direction. This domain magnetic moment results from the alignment of the moments of the individual atoms in the domain. The overall magnetization of the macroscopic material is the sum of the contributions of all the domains contained in the material.

When the magnetic moments of the individual domains are randomly directed, no net magnetization (**M**) occurs. Full magnetization (saturation) occurs when the magnetic moments of all the domains are aligned in the same direction. A range of magnetizations for the macroscopic material, between the extremes of zero and saturation, can be achieved depending on the degree of alignment of the domains in the material.

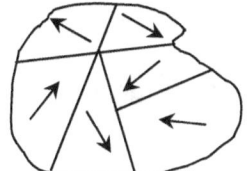
Random alignment
No net magnetization

Partial alignment
Partial magnetization

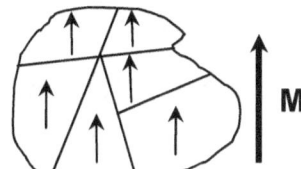
Full alignment
Saturation

Ferromagnetic domains can be as large as a few millimeters or as small as a micrometer and they can vary in shape. If a magnetic field (**H**) is applied to the material, a net magnetization is achieved due to either a realignment of the domains in the direction of the applied field or due to a change in the size of the domains favorable to a net magnetization in the direction of the applied field.

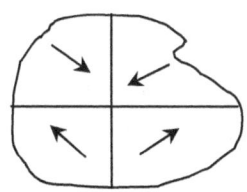
Random alignment
No net magnetization

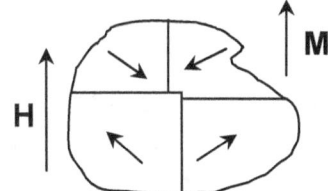
Magnetization by movement of domain boundaries

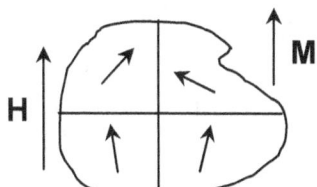
Magnetization by alignment of domains

The level of magnetization in the material depends on the strength of the applied field and can range from a small amount all the way to saturation. When the applied field is removed, some magnetization may remain however. Residual magnetization implies that the present state of the material, magnetically, depends on its past magnetic history (hysteresis). A so-called hysteresis loop describes the magnetization of the material in the presence of an applied field.

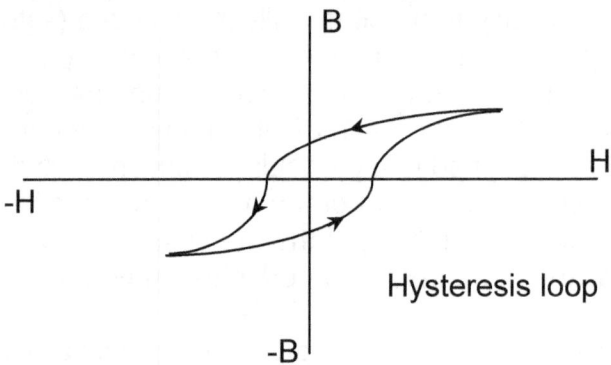

Hysteresis loop

The above depiction is for one possible hysteresis loop for some particular material. There are, in fact, an infinite number of loops, depending on the range of applied field strengths. The shape depends on both present and past values of the applied field as well as the tendency of the ferromagnetic domains to return to their original (possibly random) alignment when the applied field is decreased or eliminated.

Skill 9.2 Determining the strength and orientation of the magnetic field near a current-carrying wire

Conductors through which electrical currents travel will produce magnetic fields: The magnetic field *dB* induced at a distance *r* by an element of current *Idl* flowing through a wire element of length *dl* is given by the **Biot-Savart** law

$$dB = \frac{\mu_0}{4\pi} \frac{Idl \times \hat{r}}{r^2}$$

where μ_0 is a constant known as the permeability of free space and $\hat{r}$ is the unit vector pointing from the current element to the point where the magnetic field is calculated.

An alternate statement of this law is **Ampere's law** according to which the line integral of *B.dl* around any closed path enclosing a steady current *I* is given by

$$\oint_C B \cdot dl = \mu_0 I$$

The basis of this phenomenon is the same no matter what the shape of the conductor, but we will consider three common situations:

Straight Wire

Around a current-carrying straight wire, the magnetic field lines form concentric circles around the wire. The direction of the magnetic field is given by the right-hand rule: When the thumb of the right hand points in the direction of the current, the fingers curl around the wire in the direction of the magnetic field. Note the direction of the current and magnetic field in the diagram.

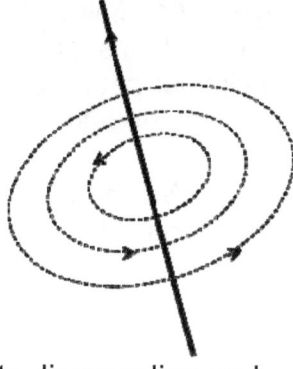

To find the magnetic field of an infinitely long (allowing us to disregarding end effects) we apply Ampere's Law to a circular path at a distance r around the wire:

$$B = \frac{\mu_0 I}{2\pi r}$$

where μ_0=the permeability of free space ($4\pi \times 10^{-7}$ T·m/A)
I=current
r=distance from the wire

Skill 9.3 Solving problems involving charged particles moving through a magnetic field

The magnetic force exerted on a charge moving in a magnetic field depends on the size and velocity of the charge as well as the magnitude of the magnetic field. One important fact to remember is that only the velocity of the charge in a direction perpendicular to the magnetic field will affect the force exerted. Therefore, a charge moving parallel to the magnetic field will have no force acting upon it whereas a charge will feel the greatest force when moving perpendicular to the magnetic field.

The direction of the magnetic force is always at a right angle to the plane formed by the velocity vector v and the magnetic field B and is given by applying the right hand rule - if the fingers of the right hand are curled in a way that seems to rotate the v vector into the B vector, the thumb points in the direction of the force. The magnitude of the force is equal to the cross product of the velocity of the charge with the magnetic field multiplied by the magnitude of the charge.

$$F=q \, (v \times B) \quad or \quad F=q \, v \, B\sin(\theta)$$

Where θ is the angle formed between the vectors of velocity of the charge and direction of magnetic field.

Problem: Assuming we have a particle of 1×10^{-6} kg that has a charge of -8 coulombs that is moving perpendicular to a magnetic field in a clockwise direction on a circular path with a radius of 2 m and a speed of 2000 m/s, let's determine the magnitude and direction of the magnetic field acting upon it.

Solution: We know the mass, charge, speed, and path radius of the charged particle. Combining the equation above with the equation for centripetal force we get

$$qvB = \frac{mv^2}{r} \quad \text{or} \quad B = \frac{mv}{qr}$$

Thus B= $(1 \times 10^{-6}$ kg$)$ $(2000$m/s$)$ / $(-8$ C$)(2$ m$)$ = 1.25×10^{-4} Tesla

Since the particle is moving in a clockwise direction, we use the right hand rule and point our fingers clockwise along a circular path in the plane of the paper while pointing the thumb towards the center in the direction of the centripetal force. This requires the fingers to curl in a way that indicates that the magnetic field is pointing out of the page. However, since the particle has a negative charge we must reverse the final direction of the magnetic field into the page.

A mass spectrometer measures the mass to charge ratio of ions using a setup similar to the one described above. m/q is determined by measuring the path radius of particles of known velocity moving in a known magnetic field.

Skill 9.4 Demonstrating knowledge of direction and relative magnitude of an induced EMF in a conductor

When the magnetic flux through a coil is changed, a voltage is produced which is known as induced electromagnetic force. Magnetic flux is a term used to describe the number of magnetic fields lines that pass through an area and is described by the equation:

$$\Phi = B A \cos\theta$$

Where Φ is the angle between the magnetic field B, and the normal to the plane of the coil of area A

By changing any of these three inputs, magnetic field, area of coil, or angle between field and coil, the flux will change and an EMF can be induced. The speed at which these changes occur also affects the magnitude of the EMF, as a more rapid transition generates more EMF than a gradual one. This is described by **Faraday's law** of induction:

$$\varepsilon = -N \Delta\Phi / \Delta t$$

where ε is emf induced, N is the number of loops in a coil, t is time, and Φ is magnetic flux

The negative sign signifies **Lenz's law** which states that induced emf in a coil acts to oppose any change in magnetic flux. Thus the current flows in a way that creates a magnetic field in the direction opposing the change in flux. See the next section for the right-hand rule that determines the direction of the induced current.

Consider a coil lying flat on the page with a square cross section that is 10 cm by 5 cm. The coil consists of 10 loops and has a magnetic field of 0.5 T passing through it coming out of the page. Let's find the induced EMF when the magnetic field is changed to 0.8 T in 2 seconds.

First, let's find the initial magnetic flux: Φ_i
$\Phi_i = BA \cos \theta = (.5 \text{ T})(.05 \text{ m})(.1 \text{ m}) \cos 0° = 0.0025 \text{ T m}^2$

And the final magnetic flux: Φ_f
$\Phi_f = BA \cos \theta = (0.8 \text{ T})(.05 \text{ m})(.1 \text{ m}) \cos 0° = 0.004 \text{ T m}^2$

The induced emf is calculated then by
$\varepsilon = -N \Delta\Phi / \Delta t = -10 (.004 \text{ T m}^2 - .0025 \text{ T m}^2) / 2 \text{ s} = -0.0075 \text{ volts}$.

To determine the direction the current flows in the coil we need to apply the right hand rule and Lenz's law. The magnetic flux is being increased out of the page, with your thumb pointing up the fingers are coiling counterclockwise. However, Lenz's law tells us the current will oppose the change in flux so the current in the coil will be flowing clockwise.

The negative sign in Faraday's Law (described in the previous section) leads to Lenz's law which states that the induced current must produce a magnetic field that opposes the change in the applied magnetic field. This is simply an expression of the conservation of energy.

The right- or left-hand rule applies only in the case of a particular current convention: the right-hand rule is typically applied in the case of the positive current convention (i.e., the direction of flow of positive charge). Since positive and negative current are complementary, however, the rules for negative current simply involve using the left hand instead of the right hand. For simplicity, only the right-hand rule will be discussed explicitly.

The right-hand rule states that if the fingers of the right hand are curled in the direction of the positive current flow, the resulting magnetic flux is in the direction of the extended thumb. According to Lenz's law, when the magnetic flux through the surface bounded by the conductor increases, the induced current, $i_{induced}$, must produce a magnetic flux that is in the opposite direction to that of the applied flux. This is illustrated below.

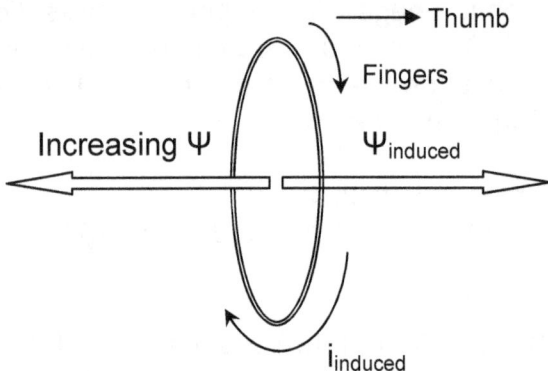

When the applied magnetic flux through the surface decreases, the direction of the induced flux must likewise coincide with the direction of the applied flux.

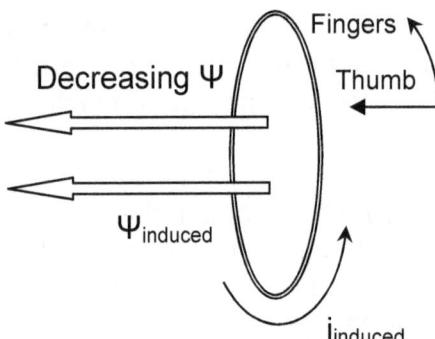

Thus, the induced magnetic flux must oppose the change in the applied flux. The direction of the induced positive current is the same as the direction of the curled fingers of the right hand when the thumb is extended in the direction of the induced magnetic flux. If negative current is of interest (for example, the flow of electrons), then the left-hand must be substituted in place of the right hand; the thumb follows the same rules, but the direction of the curled fingers indicates the direction of negative current flow.

Skill 9.5 Analyzing applications of electromagnetism

Electromagnetism is the foundation for a vast number of modern technologies ranging from computers to communications equipment. More mundane technologies such as motors and generators are also based upon the principles of electromagnetism. The particular understanding of electrodynamics can either be in terms of quantum mechanics (quantum electrodynamics) or classical electrodynamics, depending on the type of phenomenon being analyzed. In classical electrodynamics, which is a sufficient approximation for most situations, the electric field, resulting from electric charge, and the magnetic field, resulting from moving charges, are the parameters of interest and are related through Maxwell's equations.

Motors

Electric motors are found in many common appliances such as fans and washing machines. The operation of a motor is based on the principle that a magnetic field exerts a force on a current carrying conductor. This force is essentially due to the fact that the current carrying conductor itself generates a magnetic field; the basic principle that governs the behavior of an electromagnet. In a motor, this idea is used to convert **electrical energy into mechanical energy**, most commonly rotational energy. Thus the components of the simplest motors must include a strong magnet and a current-carrying coil placed in the magnetic field in such a way that the force on it causes it to rotate.

A typical motor is composed of a stationary portion, called the stator, and a rotating (or moving) portion, called the rotor. Coils of wire that serve as electromagnets are wound on the armature, which can be either the stator or the rotor, and are powered by an electric source. Motors use electric current to generate a magnetic field around an electromagnet, which results in a rotational force due to the presence of an external magnetic field (either from permanent magnets or electromagnets). The designs of various motors can differ dramatically, but the general principles of electromagnetism that describe their operation are generally the same.

Generators

Generators are in effect "reverse motors". They exploit electromagnetic induction to generate electricity. Thus, if a coil of wire is rotated in a magnetic field, an alternating EMF is produced which allows current to flow. Any number of energy sources can be used to rotate the coil, including combustion, nuclear fission, flowing water or other sources. Therefore, generators are devices that convert **mechanical or other forms of energy into electrical energy**.

Meters

A number of different types of meters use electromagnetism or are designed to measure certain electromagnetic parameters. For example, older forms of ammeters (galvanometers), when supplied with a current, provided a measurement through the deflection of a spring-loaded needle. A coil connected to the needle acted as an electromagnet which, in the presence of a permanent magnetic field, would be deflected in the same manner as a rotor, as mentioned previously. The strength of the electromagnet, and thus the extent of the deflection, is proportional to the current. Further, the spring limits the deflection in such a manner that a reasonably accurate measurement of the current is provided.

Magnetic Media

Although the cassette tape has fallen out of favor with popular culture, magnetic media are still widely used for information storage. Magnetic strips on the back of credit and identification cards, computer hard drive disks and magnetic tapes (such as those contained in cassettes) are all examples of magnetic media. The principle underlying magnetic media is the sequential magnetization of a region of the medium. A special head is able to detect spatial magnetic fluctuations in the medium which are then converted into an electrical signal. The head is often able to "write" to the medium as well. The electrical signal from the medium can be converted into sound or video, as with the video or audio cassette player, or it can be digitized for use in a computer.

Transformers

Transformers and generators have critical functions in the production and use of electrical power. A generator is essentially a transducer that converts one form of energy, such as mechanical or heat energy, into electrical energy. It uses electromagnetic induction (see **Skill 9.4**) to generate electricity. The simplest way to implement a generator is to rotate a wire loop in a permanent magnetic field.

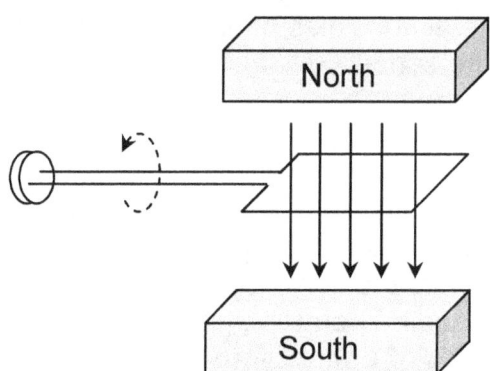

Depending on the strength of the magnetic field and the rate of rotation of the loop, a voltage of a specific amplitude can be produced. Additionally, the rotation of the wire loop means that the magnetic flux is at times increasing and at times decreasing, resulting in a sinusoidal voltage with a frequency equal to the rotation frequency of the loop. Electricity supplied by public utilities, for example, is 120 V at a frequency of 60 Hz. The source of energy for the wire loop rotator (turbine) can be a dammed river or an engine run by combustion, a nuclear reaction or another process. The energy extracted from these sources (either as mechanical or heat energy) is transferred to electrical energy through the use of the generator.

Transformers fulfill another purpose in the context of electrical power by either stepping up or stepping down the voltage (and, concomitantly, stepping down or stepping up the current). They do this by magnetically coupling two circuits together. This allows the transfer of energy between these two circuits without requiring motion. Typically, a transformer consists of a couple of coils and a magnetic core. A changing voltage applied to one coil (the primary inductor) creates a flux in the magnetic core, which induces voltage in the other coil (the secondary inductor).

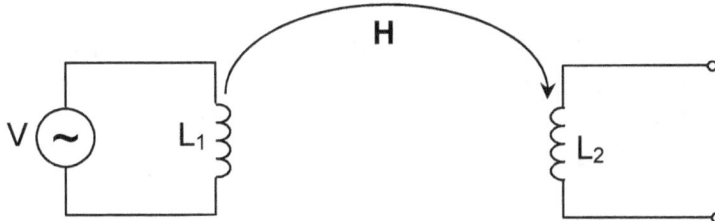

One of the simplest ways to control the flux linkage between the coils is by appropriately increasing or decreasing the number of turns in the inductors relative to one another. In the ideal case, where all the flux produced by the first inductor (L_1 above) is linked to the second inductor (L_2 above), the voltage across L_2 (V_t) is equal to the voltage across L_1 (V) multiplied by the ratio of the number of turns in L_2 (N_2) to the number of turns in L_1 (N_1).

$$V_t = V \frac{N_2}{N_1}$$

Thus, if there are twice as many turns in L_2 as in L_1, twice the flux is linked to L_2 and the voltage is doubled. In reality, however, not all the flux is linked to the second inductor. The use of ferromagnetic cores, especially if a loop of ferromagnetic material is used such that it serves as the core for both inductors, increases the flux linkage and brings the transformer closer to the ideal case. All transformers operate on this simple principle though they range in size and function from those in tiny microphones to those that connect the components of the US power grid.

Skill 9.6 Demonstrating knowledge of the generation of electromagnetic waves and their applications

Waves are simply disturbances that propagate through space and time. Most waves must propagate through a medium, which may be solid, liquid, or gas. However, some electromagnetic waves do not require a medium and can move through a vacuum. Most waves transfer energy from their source to their destination, typically without actually transferring molecules of the medium through which they travel. The individual particles of the medium are temporarily displaced but eventually return to their equilibrium positions. As the particles oscillate, they transfer energy and momentum to a neighboring particle which, in turn, moves back and forth around its equilibrium position while transferring energy and momentum to another particle and so on.

The characteristics of electromagnetic radiation vary depending on the frequency of the radiation and the properties of the material through which it is propagating. Virtually innumerable applications have been devised for various frequency bands, leading to the development of a number of interesting and practical devices and inventions.

A common application of the microwave portion of the electromagnetic spectrum is the microwave oven. Since water molecules have a local charge imbalance, they each manifest a dipole moment. Microwave radiation operates at a wavelength such that incidence on water molecules imparts oscillating motion because of the dipole moments. This motion causes the temperature of the water to rise, thus cooking the food.

Another useful application of the electromagnetic spectrum is remote communications. A number of different frequency bands can be exploited, each with its own characteristics. Since oscillation of charges produces electromagnetic waves that propagate away from the source, antennas that are supplied with an electrical signal can be used to send or receive information through modulation of the waves. Two common modulation schemes are amplitude modulation (AM), for which the amplitude of a sine wave at a given frequency is varied, and frequency modulation (FM), for which the frequency of a sine wave (with a specific central frequency) is modulated. In both cases, these modulation schemes use a band of frequencies (bandwidth) in accordance with Fourier theory. Thus, a broadcast radio station is actually allocated a range of frequencies around a single central frequency (50 kHz around 100.5 MHz, for example). A virtually unlimited range of frequencies can be used for communication although some are more suited to certain situations than others. For example, line-of-sight communication is amenable to the use of a visible (or non-visible) laser as well as microwaves which are commonly used for cellular phones.

In cases where visible light is dim or absent, infrared detectors or video equipment can be used to produce images of scenes that appear dark to the unaided eye. Depending on their temperature, objects radiate infrared waves which can be used to produce false-color images of an otherwise dark environment.

Another method of imaging is the use of X-rays. These high-energy (i.e., high frequency) waves are able to penetrate certain materials, such as flesh, but are unable to penetrate other materials, such as bone or metal. Thus, X-rays may be used to visualize the skeletal structure of a human or animal without an invasive procedure or they may be used to detect hidden metallic items, such as various forms of contraband.

An advanced form of imaging is holography, which allows recording of a three-dimensional image, rather than just a two-dimensional image. Holography uses the coherent light produced by lasers to imprint an image on a photographic plate; in addition to amplitude information, which is all that is recorded in standard photography, holography also includes phase information, thus allowing capture of a three-dimensional image.

The broad range of frequencies radiated by the Sun can be collected in a number of ways to beneficially use or convert the incident energy. So-called solar panels can be used to convert electromagnetic radiation into electricity for powering devices or for storage in batteries. Also, the heat produced as a result of absorption of solar radiation can be used, for example, to heat a house. An appropriate system to collect the radiation, convert it to heat and distribute it throughout a house or other structure is one particular use of this phenomenon.

COMPETENCY 10.0 UNDERSTAND THE PROPERTIES OF ELECTIC CIRCUITS

Skill 10.1 Interpreting simple schematic diagrams of DC circuits

The two most important elements in simple circuits are resistors and capacitors. Often resistors and capacitors are used together in series or parallel. Two components are in series if one end of the first element is connected to one end of the second component. The components are in parallel if both ends of one element are connected to the corresponding ends of another. A series circuit has a single path for current flow through all of its elements. A parallel circuit is one that requires more than one path for current flow in order to reach all of the circuit elements.

Below is a diagram demonstrating a simple circuit with resistors in parallel (on right) and in series (on left). Note the symbols used for a battery (noted V) and the resistors (noted R).

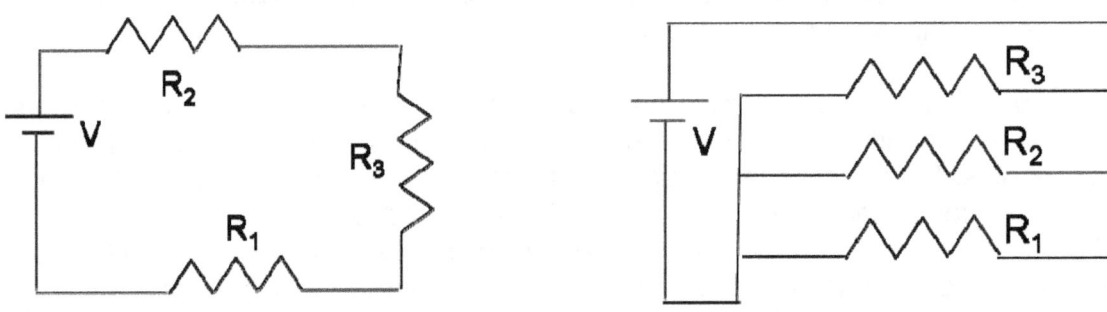

Thus, when the resistors are placed in series, the current through each one will be the same. When they are placed in parallel, the voltage through each one will be the same. To understand basic circuitry, it is important to master the rules by which the equivalent resistance (R_{eq}) or capacitance (C_{eq}) can be calculated from a number of resistors or capacitors:

Resistors in parallel: $$\frac{1}{R_{eq}} = \frac{1}{R_1} + \frac{1}{R_2} + \cdots + \frac{1}{R_n}$$

Resistors in series: $$R_{eq} = R_1 + R_2 + \cdots + R_n$$

Capacitors in parallel: $$C_{eq} = C_1 + C_2 + \cdots + C_n$$

Capacitors in series: $$\frac{1}{C_{eq}} = \frac{1}{C_1} + \frac{1}{C_2} + \cdots + \frac{1}{C_n}$$

Skill 10.2 Analyzing series and parallel circuits using Ohm's law

First review the circuit diagrams and rules for analyzing capacitors and resistors in circuits in **Skill 10.1**.

Ohm's Law is the most important tool we posses to analyze electrical circuits. Ohm's Law states that the current passing through a conductor is directly proportional to the voltage drop and inversely proportional to the resistance of the conductor. Stated mathematically, this is:

$$V = IR$$

Let's now examine a sample problem involving resistors in parallel.

Problem:
The circuit diagram at right shows three resistors connected to a battery in series. A current of 1.0A flows through the circuit in the direction shown. It is known that the equivalent resistance of this circuit is 25 Ω. What is the total voltage supplied by the battery?

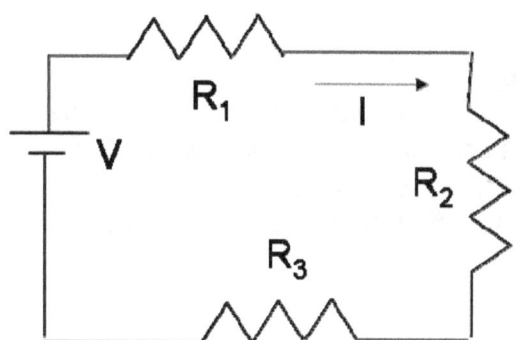

Solution:

To determine the battery's voltage, we simply apply Ohm's Law:

$$V = IR = 1.0A \times 25\Omega = 25V$$

Skill 10.3 Demonstrating knowledge of energy conservation in simple circuits

Kirchoff's Laws are a pair of laws that apply to conservation of charge and energy in circuits and were developed by Gustav Kirchoff.

Kirchoff's Current Law: At any point in a circuit where charge density is constant, the sum of currents flowing toward the point must be equal to the sum of currents flowing away from that point.

Kirchoff's Voltage Law: The sum of the electrical potential differences around a circuit must be zero.

While these statements may seem rather simple, they can be very useful in analyzing DC circuits, those involving constant circuit voltages and currents.

Problem:
The circuit diagram at right shows three resistors connected to a battery in series. A current of 1.0 A is generated by the battery. The potential drop across R₁, R₂, and R₃ are 5V, 6V, and 10V. What is the total voltage supplied by the battery?

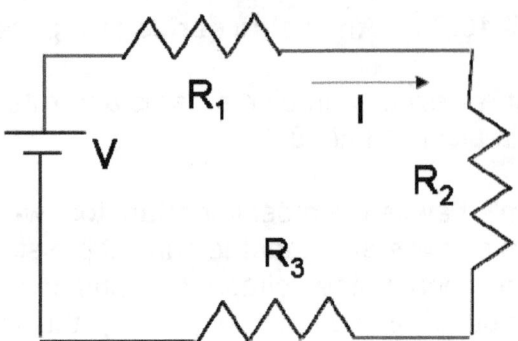

Solution:
Kirchoff's Voltage Law tells us that the total voltage supplied by the battery must be equal to the total voltage drop across the circuit. Therefore:

$$V_{battery} = V_{R_1} + V_{R_2} + V_{R_3} = 5V + 6V + 10V = 21V$$

Problem:
The circuit diagram at right shows three resistors wired in parallel with a 12V battery. The resistances of R₁, R₂, and R₃ are 4 Ω, 5 Ω, and 6 Ω, respectively. What is the total current?

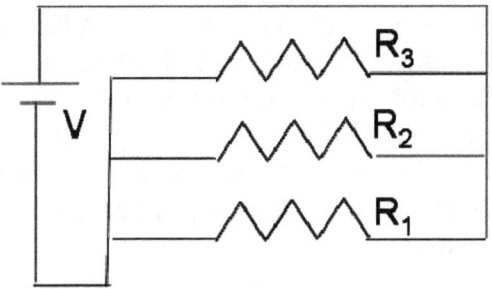

Solution:
This is a more complicated problem. Because the resistors are wired in parallel, we know that the voltage entering each resistor must be the same and equal to that supplied by the battery. We can combine this knowledge with Ohm's Law (see next section) to determine the current across each resistor:

$$I_1 = \frac{V_1}{R_1} = \frac{12V}{4\Omega} = 3A$$

$$I_2 = \frac{V_2}{R_2} = \frac{12V}{5\Omega} = 2.4A$$

$$I_3 = \frac{V_3}{R_3} = \frac{12V}{6\Omega} = 2A$$

Finally, we use Kirchoff's Current Law to find the total current:

$$I = I_1 + I_2 + I_3 = 3A + 2.4A + 2A = 7.4A$$

Skill 10.4 Differentiating between alternating and direct current circuits

Alternating current (AC) is a type of electrical current with cyclically varying magnitude and direction. This is differentiated from direct current (DC), which has constant direction. AC is the type of current delivered to businesses and residences.

Direct Current

Direct current is the simple, constant flow of electrical charge (electrons). It was first utilized by Thomas Edison in the 1880s to power a light bulb. By 1887, numerous DC power plants existed across the United States. However, DC power, in which electrons flow from a negative to a positive end, had it limitations. The current tends to lose power over relatively short distances. Many DC power plants struggled with the fact that they had a significant loss of power on the line within a mile of the plant.

Today, direct current is used, but in limited applications. DC is used for many low voltage applications. Battery powered devices are powered by direct current. Direct current is also used by commuter trains and solar cells.

Alternating Current

Westinghouse developed an alternating current in response to the problems associated with direct current. Unlike direct current, alternating current is a much more efficient transmission of power and can travel long distances without losing strength.

AC changes both direction and magnitude due to the generator. As the armature of the generator changes direction, so does the direction of the current. An alternating current will change direction 50 to 60 times per second as the charge at the end of the wire alternates between a positive charge and a negative charge.

The voltage of AC can easily be stepped up or stepped down through the use of transformers. Therefore, high voltage can be sent through power lines, and stepped down before entering homes or businesses (higher voltage is more efficient). In addition, alternating current is easy to change to direct current for use in household items. Alternating current is used to power homes, computers, televisions, etc.

COMPETENCY 11.0 UNDERSTAND THE BASIC PROCESSES OF ATOMIC AND NUCLEAR PHYSICS

Skill 11.1 Analyzing differences between fission and fusion and the applications of each

One consequence of the principle of mass-energy equivalence is that mass may be transformed into other types of energy and vice versa. An example of the conversion of mass into energy is the **binding energy** of an atomic nucleus. A deuteron, for instance, has a mass that is less than the sum of the masses its constituent parts, a neutron and a proton. The mass difference is the binding energy that holds the deuteron together.

Problem: The mass of a proton is 1.6726×10^{-27} Kg, the mass of a neutron is 1.6749×10^{-27} Kg, and the mass of a deuteron is 3.3436×10^{-27} Kg. What is the binding energy of a deuteron?

Solution: The difference in mass between the deuteron and its constituents $\Delta m =$

$1.6726 \times 10^{-27} + 1.6749 \times 10^{-27} - 3.3436 \times 10^{-27} = 0.0039 \times 10^{-27}$ Kg

Thus binding energy = $\Delta mc^2 = 3.51 \times 10^{-13} J$

Nuclear **fusion** involves the joining of several nuclei to form one heavier nucleus. Nuclear **fission** is the reverse of fusion, in that it is the splitting of a nucleus to form multiple lighter nuclei. Depending on the weight of the nuclei involved, both fission and fusion may result in either the absorption or release of energy. Iron and nickel have the largest binding energies per nucleon and so are the most stable nuclei. Thus, *fusion* releases energy when the two nuclei are lighter than iron or nickel and *fission* releases energy when the two nuclei are heavier than iron or nickel. Conversely, fusion will absorb energy when the nuclei are heavier and fission will absorb energy when the nuclei are lighter.

Nuclear fusion is common in nature and is the mechanism by which new natural elements are created. Fusion reactions power the stars and (energy absorbing) fusion of heavy elements occurs in supernova explosions. Despite the fact that significant energy is required to trigger the fusion of two nuclei (to overcome the electrostatic repulsion between the positively charged protons), the reaction can be self-sustaining because the energy released by the fusion of two light nuclei is greater than that required to force them together. Fusion is typically much harder to control that fission and so it is not used for power generation though fusion reactions are used to drive hydrogen bombs.

Nuclear fission is unique in that it can be harnessed for a variety of applications. This is done via the use of a chain reaction initiated by the bombardment of certain isotopes with free neutrons. When a nucleus is struck by a free neutron, it splits into smaller nuclei and also produces free neutrons, gamma rays, and alpha and beta particles. The free neutrons can then go onto interact with other nuclei and perpetuate the fission reaction. Isotopes, such as ^{235}U and ^{239}P that sustain the chain reaction are known as fissile and used for nuclear fuel. Because fission can be controlled via chain reaction, it is used in nuclear power generation. Uncontrolled fission reactions are also used in nuclear weapons, including the atomic bombs developed during the Manhattan Project and exploded over Hiroshima and Nagasaki in 1945.

Though it is currently in use in many locations, nuclear fission for power generation remains somewhat controversial. The amount of available energy per pound in nuclear fuel is millions of times that in fossil fuels. Additionally, nuclear power generation does not produce the air and water pollutants that are problematic byproducts of fossil fuel combustion. The currently used fission reactions, however, do produce radioactive waste that must be contained for thousands of years.

Nuclear reactions can consist of simple radioactive decay, fission, fusion, and other nuclear processes. In all cases, both **mass-energy and charge must be conserved**. Below we take a closer look at how this is so in the case of alpha and beta decay.

$$^{238}U \rightarrow {}^{234}Th + \alpha$$

The isotope of uranium in the above reaction weighs 238 Daltons. Because it is uranium, it has 92 protons, meaning it must have 146 neutrons. When it undergoes alpha decay it loses 2 protons and 2 neutrons. The alpha particle weighs 4 Daltons and the nucleus that has undergone decay weighs 234 Daltons. Thus mass is conserved. The decayed nucleus will have a charge reduced by that of 2 protons following the decay. However, the emitted alpha particle carries the additional charge due to its 2 protons. Thus both charge and mass are conserved over all.

$$^{137}_{55}Cs \rightarrow {}^{137}_{56}Ba + e^- + \bar{v}_e$$

This isotope of cesium weighs 137 Daltons and, like all cesium isotopes, it has 55 protons. When it undergoes beta minus decay, a neutron is converted to a proton and an electron and an anti-neutrino are lost. The total mass-energy of the system is conserved since the difference in mass between a neutron and an electron plus proton is balanced by the energy of the emitted electron and the anti-neutrino. In beta minus decay a neutron, with no charge, is split into a positively charged proton and a negatively charged electron. Thus the conservation of charge is satisfied. The electron is emitted, while the proton remains in the nucleus. With one extra proton, the nucleus is now a barium isotope.

$$^{22}_{11}Na \rightarrow ^{22}_{10}Ne + e^+ + v_e$$

In order to conserve mass-energy of the system, beta plus decay cannot occur in isolation but only in a nucleus since the mass of a neutron is greater than the mass of a proton plus electron. The difference in binding energy of the mother and daughter nucleus provides the additional energy needed for the reaction to go through. Charge is conserved when a positively charged proton is converted into a positively charged positron. With one fewer proton, the decayed nucleus is now a neon isotope weighing 22 Daltons.

Problem: Uranium 235 is used as a nuclear fuel in a chain reaction. The reaction is initiated by a single neutron and produces barium 141, an unknown isotope, and 3 neutrons that can go on to propagate the chain reaction. Determine the unknown isotope. Assume that the kinetic energies and the energy released in the reaction is negligible compared to the masses of the isotopes produced.

Solution: We can begin by writing out the reaction, leaving open the unknown isotope X.

$$^{235}_{92}U + ^1_0n \rightarrow ^{141}_{56}Ba + X + 3^1_0n$$

We begin with the charge balance; since the neutron has no charge, the unknown isotope must have 36 protons. Consulting a periodic table, we see that this will mean it is a Krypton isotope. Now we can balance the mass. Since the original nucleus weighed 235 Daltons and one neutron was added to it, the total mass of the resultant nuclei must be 236. So, we can simple subtract the weight of the barium isotope and the 3 new neutrons to find the unknown weight:

236-141-3=92

Thus our unknown isotope is krypton 92, making the balanced equation:

$$^{235}_{92}U + ^{1}_{0}n \rightarrow ^{141}_{56}Ba + ^{92}_{36}Kr + 3^{1}_{0}n$$

Skill 11.2   Demonstrating knowledge of models of nuclear and subatomic structures and behaviors

Some nuclei are unstable and emit particles and electromagnetic radiation. These emissions from the nucleus are known as **radioactivity.** It is often found that some isotopes of an element are radioactive, typically the ones with an excess of neutrons in the nucleus. The unstable isotopes are known as **radioisotopes**; and the nuclear reactions that spontaneously alter them are known as **radioactive decay.** Particles commonly involved in nuclear reactions are listed in the following table:

Particle	Neutron	Proton	Electron	Alpha particle	Beta particle	Gamma rays
Symbol	$^{1}_{0}n$	$^{1}_{1}p$ or $^{1}_{1}H$	$^{0}_{-1}e$	$^{4}_{2}\alpha$ or $^{4}_{2}He$	$^{0}_{-1}\beta$ or $^{0}_{-1}e$	$^{0}_{0}\gamma$

Artificial or **induced radioactivity** is the production of radioactive isotopes by bombarding an element with high velocity particles such as neutrons.

In **alpha decay**, an atom emits an alpha particle. An alpha particle contains two protons and two neutrons. This makes it identical to a helium nucleus and so an alpha particle may be written as He^{2+} or it can be denoted using the Greek letter α. Because a nucleus decaying through alpha radiation loses protons and neutrons, the mass of the atom loses about 4 Daltons and it actually becomes a different element (transmutation). For instance:

$$^{238}U \rightarrow ^{234}Th + \alpha$$

Radioactive heavy nuclei including uranium and radium typically decay by emitting alpha particles. The alpha decay often leaves the nucleus in an excited state with the extra energy subsequently removed by gamma radiation. The energy of alpha particles can be readily absorbed by skin or air and so alpha decaying substances are only harmful to living things if they are introduced internally.

Like alpha decay, **beta decay** involves emission of a particle. In this case, though, it is a beta particle, which is either an electron or positron. Note that a positron is the antimatter equivalent of an electron and so these particles are often denoted β⁻ and β⁺. Beta plus and minus decay occur via roughly opposite paths. In beta minus decay, a neutron is converted in a proton (specifically, a down quark is converted to an up quark), an electron and an anti-neutrino; the latter two are emitted. In beta plus decay, on the other hand, a proton is converted to a neutron, a positron, and a neutrino; again, the latter two are emitted. As in alpha decay, a nucleus undergoing beta decay is transmuted into a different element because the number of protons is altered. However, because the total number of nucleons remains unchanged, the atomic mass remains the same (note, that the neutron is actually slightly heavier than a proton so mass is gained during beta plus decay). So examples of beta decay would be:

$$^{137}_{55}Cs \rightarrow ^{137}_{56}Ba + e^- + \bar{v}_e \quad \text{(beta minus decay)}$$

$$^{22}_{11}Na \rightarrow ^{22}_{10}Ne + e^+ + v_e \quad \text{(beta plus decay)}$$

Beta decaying isotopes, such as Strontium 90, are commonly used in cancer treatment. These particles are better able to penetrate skin than alpha particles and so exposure to larger amounts of beta particles poses a risk to all living things.

Gamma radiation is quite different from alpha and beta decay in that it does not involve the emission of nucleon-containing particles or the transmutation of elements. Rather, gamma-ray photons are emitted during gamma decay. These gamma rays are a specific form of electromagnetic radiation that results from certain sub-atomic particle contacts. For instance, electron-positron annihilation leads to the emission of gamma rays. More commonly, though, gamma rays are emitted by nuclei left in an excited state following alpha or beta decay. Thus, gamma decay lowers the energy level of a nucleus, but does not change its atomic mass or charge. The high energy content of gamma rays, coupled with their ability to penetrate dense materials, make them a serious risk to living things.

Skill 11.3 Demonstrating knowledge of the half-life of radioactive isotopes

While the radioactive decay of an individual atom is impossible to predict, a mass of radioactive material will decay at a specific rate. Radioactive isotopes exhibit exponential decay and we can express this decay in a useful equation as follows:

$$A = A_0 e^{kt}$$

Where A is the amount of radioactive material remaining after time t, A_0 is the original amount of radioactive material, t is the elapsed time, and k is the unique activity of the radioactive material. Note that k is unique to each radioactive isotope and it specifies how quickly the material decays. Sometimes it is convenient to express the rate of decay as half-life. A half-life is the time needed for half a given mass of radioactive material to decay. Thus, after one half-life, 50% of an original mass will have decayed, after two half lives, 75% will have decayed and so on.

Let's examine a sample problem related to radioactive decay.

Problem: Radiocarbon dating has been used extensively to determine the age of fossilized organic remains. It is based on the fact that while most of the carbon atoms in living things is ^{12}C, a small percentage is ^{14}C. Since ^{14}C is a radioactive isotope, it is lost from a fossilized specimen at a specific rate following the death of an organism. The original and current mass of ^{14}C can be inferred from the relative amount of ^{12}C. So, if the half-life of ^{14}C is 5730 years and a specimen that originally contained 1.28 mg of ^{14}C now contains 0.10 mg, how old is the specimen?

In certain problems, we may be simply provided with the activity, k, but in this problem we must use the information given about half-life to solve for k.

Since we know that after one half-life, 50% of the material remains radioactive, we can plug into the governing equation above:

$$A = A_0 e^{kt}$$

$$0.5 A_0 = A_0 e^{5730k}$$

$$k = (\ln(0.5))/5730 = -0.0001209$$

Having determined k, we can use this same equation again to determine how old the specimen described above must be:

$$A = A_0 e^{kt}$$

$$0.10 = 1.28 e^{-0.0001209t}$$

$$t = \frac{\ln\left(\frac{0.10}{1.28}\right)}{-0.0001209} = 21087$$

Thus, the specimen is 21,087 years old.

Note that this same equation can be used to calculate the half-life of an isotope if information regarding the decay after a given number of years were provided.

A radioactive isotope often decays into another element that is radioactive as well and continues to decay into a third element. The process continues until a stable element is reached. This chain of disintegration is known as a **decay chain** or a **nuclear disintegration series**. For example, Uranium 238 decays into Radium 226 which decays further into Radon 222.

Skill 11.4   Demonstrating knowledge of how the basic principles of quantum mechanics can be used to describe the properties of light and matter

The wave theory of light explains many different phenomena but falls short when describing effects such as **blackbody radiation** and the **photoelectric effect**.

Blackbody radiation is the characteristic radiation of an ideal blackbody, i.e. a body that absorbs all the radiation incident upon it. Theoretical calculations of the frequency distribution of this radiation using classical physics showed that the energy density of this wave should increase as frequency increases. This result agreed with experiments at shorter wavelengths but failed at large wavelengths where experiment shows that that the energy density of the radiation actually falls back to zero.

In trying to resolve this impasse and derive the spectral distribution of blackbody radiation, Max Planck proposed that an atom can absorb or emit energy only in chunks known as quanta. The energy E contained in each quantum depends on the frequency of the radiation and is given by $E = hf$ where Planck's constant $h = 6.626 \times 10^{-34} J.s = 4.136 \times 10^{-15} eV.s$. Using this quantum hypothesis, Planck was able to provide an explanation for blackbody radiation that matched experiment.

Einstein extended Planck's idea further to suggest that quantization is a fundamental property of electromagnetic radiation which consists of quanta of energy known as **photons**. The energy of each photon is hf where h is Planck's constant.

Problem: A light beam has an intensity of 2W and wavelength of 600nm. What is the energy of each photon in the beam? How many photons are emitted by the beam every second?

Solution: The energy of each photon is given by
$E = hc/\lambda = 6.626 \times 10^{-34} \times 3 \times 10^8 / (600 \times 10^{-9}) = 3.31 \times 10^{-19} J$.
The number of photons emitted each second = $2/(3.31 \times 10^{-19}) = 6.04 \times 10^{18}$.

Einstein used the photon hypothesis to explain the photoelectric effect.

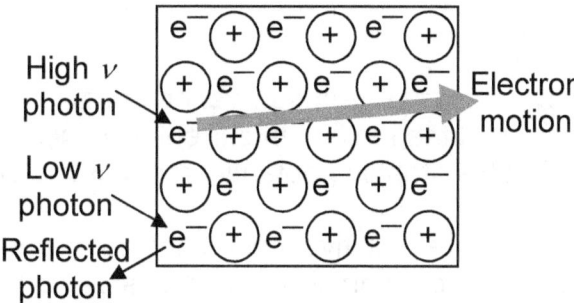

The **photoelectric effect** occurs when **light shining on a clean metal surface causes the surface to emit electrons**. The energy of an absorbed photon is transferred to an electron as shown to the right. If this energy is greater than the binding energy holding the electron close to nearby nuclei then the electron will move. A high energy (high frequency, low wavelength) photon will not only dislodge an electron from the "electron sea" of a metal but it will also impart kinetic energy to the electron, making it move rapidly. These electrons in motion will produce an electric current if a circuit is present.

When the metal surface on which light is incident is a cathode with the anode held at a higher potential V, an electric current flows in the external circuit. It is observed that current flows only for light of higher frequencies. Also there is a threshold negative potential, the **stopping potential** V_0 below which no current will flow in the circuit.

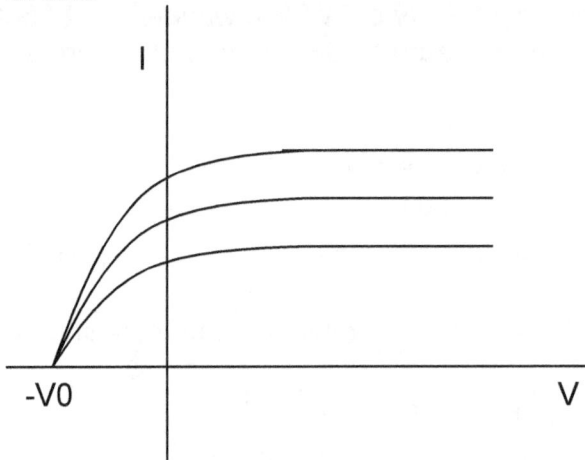

The figure displayed above shows current flow vs. potential for three different intensities of light. It shows that the maximum current flow increases with increasing light intensity but the stopping potential remains the same.

All these observations are counter-intuitive if one considers light to be a wave but may be understood in terms of light particles or photons. According to this interpretation, each photon transfers its energy to a single electron in the metal. Since the energy of a photon depends on its frequency, only a photon of higher frequency can transfer enough energy to an electron to enable it to pass the stopping potential threshold.

When V is negative, only electrons with a kinetic energy greater than |eV| can reach the anode. The maximum kinetic energy of the emitted electrons is given by eV_0. This is expressed by Einstein's photoelectric equation as

$$(\tfrac{1}{2}mv^2)_{max} = eV_0 = hf - \varphi$$

where the **work function** φ is the energy needed to release an electron from the metal and is characteristic of the metal.

Problem: The work function for potassium is 2.20eV. What is the stopping potential for light of wavelength 400nm?

Solution:
$eV_0 = hf - \varphi = hc/\lambda - \varphi = 4.136 \times 10^{-15} \times 3 \times 10^8 / (400 \times 10^{-9}) - 2.20 = 3.10 - 2.20 = 0.90\text{eV}$

Thus stopping potential $V_0 = 0.90\text{V}$

As quantum theory was developed and popularized (primarily by Max Planck and Albert Einstein), chemists and physicists began to consider how it might apply to atomic structure. Niels Bohr put forward a model of the atom in which electrons could only orbit the nucleus in circular orbitals with specific distances from the nucleus, energy levels, and angular momentums. In this model, electrons could only make instantaneous "quantum leaps" between the fixed energy levels of the various orbitals. The Bohr model of the atom was altered slightly by Arnold Sommerfeld in 1916 to reflect the fact that the orbitals were elliptical instead of round.

Though the Bohr model is still thought to be largely correct, it was discovered that electrons do not truly occupy neat, cleanly defined orbitals. Rather, they exist as more of an "electron cloud." The work of Louis de Broglie, Erwin Schrödinger, and Werner Heisenberg showed that an electron can actually be located at any distance from the nucleus. However, we can find the *probability* that the electrons exists at given energy levels (i.e., in particular orbitals) and those probabilities will show that the electrons are most frequently organized within the orbitals originally described in the Bohr model.

The quantum structure of the atom describes electrons in discrete energy levels surrounding the nucleus. When an electron moves from a high energy orbital to a lower energy orbital, a quantum of electromagnetic radiation is emitted, and for an electron to move from a low energy to a higher energy level, a quantum of radiation must be absorbed. The particle that carries this electromagnetic force is called a **photon**. The quantum structure of the atom predicts that only photons corresponding to certain wavelengths of light will be emitted or absorbed by atoms. These distinct wavelengths are measured by **atomic spectroscopy**.

An electron may exist at distinct radial distances (r_n) from the nucleus. These distances are proportional to the square of the **principal quantum number**, n. For a hydrogen atom (shown at left), the proportionality constant is called the **Bohr radius** ($a_0 = 5.29 \times 10^{-11}$ m). This value is the mean distance of an electron from the nucleus at the ground state of $n = 1$. The distances of other electron shells are found by the formula:
$$r_n = a_0 n^2.$$
As $n \to \infty$, the electron is no longer part of the hydrogen atom. Ionization occurs and the atom become an H^+ ion.

A quantum of energy (ΔE) emitted from or absorbed by an electron transition is directly proportional to the frequency of radiation. The proportionality constant between them is **Planck's constant** ($h = 6.63 \times 10^{-34}$ J·s):
$$\Delta E = h\nu \quad \text{and} \quad \Delta E = \frac{hc}{\lambda}.$$

Quantum #	Radius
$n \to \infty$	$r_\infty \to \infty$
$\vdots$	$\vdots$
$n = 5$	$r_5 = 25a_0$
$n = 4$	$r_4 = 16a_0$
$n = 3$	$r_3 = 9a_0$
$n = 2$	$r_2 = 4a_0$
$n = 1$	$r_1 = a_0$
$\oplus$ (H nucleus)	

An electron may exist at distinct radial distances (r_n) from the nucleus. These distances are proportional to the square of the **principal quantum number**, n. For a hydrogen atom (shown at left), the proportionality constant is called the **Bohr radius** ($a_0 = 5.29 \times 10^{-11}$ m). This value is the mean distance of an electron from the nucleus at the ground state of $n = 1$. The distances of other electron shells are found by the formula:

$$r_n = a_0 n^2.$$

As $n \to \infty$, the electron is no longer part of the hydrogen atom. Ionization occurs and the atom become an H^+ ion.

A quantum of energy (ΔE) emitted from or absorbed by an electron transition is directly proportional to the frequency of radiation. The proportionality constant between them is **Planck's constant** ($h = 6.63 \times 10^{-34}$ J·s):

$$\Delta E = h\nu \quad \text{and} \quad \Delta E = \frac{hc}{\lambda}.$$

The energy of an electron (E_n) is inversely proportional to its radius from the nucleus. For a hydrogen atom (shown below left), only the principle quantum number determines the energy of an electron by the **Rydberg constant** ($R_H = 2.18 \times 10^{-18}$ J):

Quantum #	Energy
$n \to \infty$	$E_\infty \to 0$
$\vdots$	$\vdots$
$n = 3$	$E_3 = -\dfrac{R_H}{9}$
$n = 2$	$E_2 = -\dfrac{R_H}{4}$
$n = 1$	$E_1 = -R_H$

$$E_n = -\frac{R_H}{n^2}.$$

The Rydberg constant is used to determine the energy of a photon emitted or absorbed by an electron transition from one shell to another in the H atom:

$$\Delta E = R_H \left(\frac{1}{n_{initial}^2} - \frac{1}{n_{final}^2} \right).$$

When a photon is absorbed, n_{final} is greater than $n_{initial}$, resulting in positive values corresponding to an endothermic process. Ionization occurs when sufficient energy is added for the atom to lose its electron from the ground state. This corresponds to an electron transition from $n_{initial} = 1$ to $n_{final} \to \infty$. The Rydberg constant is the energy required to ionize one atom of hydrogen. Photon emission causes negative values corresponding to an exothermic process because $n_{initial}$ is greater than n_{final}.

Planck's constant and the speed of light are often used to express the Rydberg constant in units of s^{-1} or length. The formulas below determine the photon frequency or wavelength corresponding to a given electron transition:

$$\nu_{photon} = \left(\frac{R_H}{h}\right)\left|\frac{1}{n_{initial}^2} - \frac{1}{n_{final}^2}\right| \quad \text{and} \quad \lambda_{photon} = \frac{1}{\left(\frac{R_H}{hc}\right)\left|\frac{1}{n_{initial}^2} - \frac{1}{n_{final}^2}\right|}.$$

These formulas **relate observed lines in the hydrogen spectrum to individual transitions** from one quantum state to another.

Wave – Particule Duality

The dual wave and particle nature of light has long been considered. In 1924 **Louis de Broglie** suggested that not only light but all matter, particularly electrons, may exhibit wave as well as particle behavior. He proposed that the frequency f and wavelength λ of electron waves are given by the equations

$$f = \frac{E}{h}; \lambda = \frac{h}{p}$$

where p is the momentum of the electron, E is its energy and h is Planck's constant. These are the same relations that Planck proposed for photons. Using deBroglie's equations and considering electrons as standing waves in a circular Bohr orbit, the discrete energy states of an electron could be explained and led to the same set of energy levels found by Bohr. Schrodinger developed these ideas into **wave mechanics**, a general method for finding the quantization condition for a system.

Wave-particle duality is also expressed by **Heisenberg's uncertainty principle** which places a limit on the accuracy with which one can measure the properties of a physical system. This limit is not due to the imperfections of measuring instruments or experimental methods but arises from the fundamental wave-particle duality inherent in quantum systems.

One statement of the uncertainty principle is made in terms of the position and momentum of a particle. If Δx is the uncertainty in the position of a particle in one dimension and Δp the uncertainty in its momentum in that dimension, then according to the uncertainty principle

$$\Delta x \Delta p \geq \hbar/2$$

where the reduced Planck's constant $\hbar = h/2\pi = 1.05457168 \times 10^{-34} \, J.s$

Thus if we measure the position of a particle with greater and greater accuracy, at some point the accuracy in the measurement of its momentum will begin to fall. A simple way to understand this is by considering the wave nature of a subatomic particle. If the wave has a single wavelength, then the momentum of the particle is also exactly known using the DeBroglie momentum-wavelength relationship. The position of the wave, however, extends through all space. If waves of several different wavelengths are superposed, the position of the wave becomes increasingly localized as more wavelengths are added. The increased spread in wavelength, however, then results in an increased momentum spread.

An alternate statement of the uncertainty principle may be made in terms of energy and time: $\Delta E \Delta t \geq \hbar/2$

Thus, for a particle that has a very short lifetime, the uncertainty in the determination of its energy will be large.

SUBAREA IV. CHARACTERISTICS OF SCIENCE

COMPETENCY 12.0 UNDERSTAND THE CHARACTERISTICS OF SCIENTIFIC KNOWLEDGE AND THE PROCESS OF SCIENTIFIC INQUIRY

Skill 12.1 Demonstrating knowledge of the nature, purposes, and characteristics of science

Application of the scientific method requires familiarity with certain skills that are common to all disciplines. The tools used in each case will depend on the area of study and the specific subject of study. What is common is the mode and attitude with which each skill is applied. Needless to say, uncompromising honesty and reporting of observations with as much objectivity as possible is a fundamental requirement of the scientific process.

Observing: All scientific theories and laws ultimately rest on a strong foundation of experiment. Observation, whether by looking through a microscope or by measuring with a voltmeter, is the fundamental method by which a scientist interacts with the environment to gather the needed data. Scientific observations are not just casual scrutiny but are made in the context of a rigorously planned experiment that specifies precisely what is to be observed and how. Observations must be repeated and the conditions under which they are made clearly noted in order to ensure their validity.

Hypothesizing: Hypothesizing is proposing an answer to a scientific question in order to set the parameters for experiment and to decide what is to be observed and under what conditions. A hypothesis is not a random guess but an educated conjecture based on existing theories and related experiments and a process of rigorous logical reasoning from these basics.

Ordering: For experimental or calculated data to be useful and amenable to analysis, it must be organized appropriately. How data is ordered depends on the question under investigation and the observation process. Ordering may involve prioritizing or categorizing. A data set may be ordered in multiple ways with respect to different variables in order to perform different kinds of analysis on it.

Categorizing: Categorizing is part of the process of ordering or organizing data either into known groups or by identifying new groups through review of the data. The groups may be formed in multiple dimensions, i.e. with respect to more than one variable. For example, a group of objects may be categorized by color as well as size. For some scientific experiments, categorizing may be the goal of the investigation. Categorized data is typically presented in tabular form with the category names as headings.

Comparing: Comparing equivalent quantities is one of the fundamental processes of science. In some cases an observation may be compared with a known or standard number in order to ascertain whether it meets certain criteria. In other cases, data points may be compared with each other for the purpose of prioritizing, categorizing or graphing. Before comparing one must ensure that the numbers are expressed in the same units.

Inferring: Once data has been organized, graphed and analyzed, a scientist draws conclusions or inferences based on logical reasoning from what he/she sees. An inference is generally drawn in the context of the initial hypothesis. An inference addresses whether the data disproves or supports the hypothesis and to what extent. An inference may also be drawn about some aspect of the data that was not included in the hypothesis. It may lead to the formulation of new problems and new hypotheses or provide answers to questions other than those asked in the hypothesis.

Applying: Applying is the process of connecting a theory, law or thought process to a physical situation, experimental set up or data. Not all laws are applicable to all situations. Also, a theory may be applied to data, for instance, only when it is organized in a specific way. Science requires the ability to evaluate when and how ideas and theories are applicable in specific cases.

Communicating: Communicating, both orally and by writing, is a vital part of scientific activity. Science is never done in a vacuum and theories and experiments are validated only when other scientists can reproduce them and agree with the conclusions. Also, scientific theories can be put to practical use only when others are able to understand them clearly. Thus communication, particularly with peers, is critical to the success of science.

Skill 12.2 Recognizing the dynamic nature of scientific knowledge through continual testing, revision, and the occasional rejection of existing theories

Scientific knowledge is based on a firm foundation of observation. Though mathematics and logic play a major role in defining and deducing scientific theories, ultimately even the most beautiful and intricate theory has to win the support of experiment. A single observation that contradicts an established theory can bring the whole edifice down if confirmed and reproduced. Thus scientific knowledge can never be totally certain and is always open to change based on some new evidence.

Sometimes advanced measuring devices and new equipment make it possible for scientists to detect phenomenon that no one had noted before. Nothing seemed more certain than classical Newtonian physics which explained everything from the motion of the planets to the behavior of earthly objects. At the end of the nineteenth century Lord Kelvin expressed the opinion that physics was complete except for the existence of "two small clouds"; the null result of the Michelson-Morley experiment and the failure of classical physics to predict the spectral distribution of blackbody radiation. The "two small clouds" turned out to be far more significant than Lord Kelvin could have imagined and led to the birth of relativity and quantum theory both of which totally changed the way we see the nature of reality.

If scientific knowledge is not inviolable, what keeps it from being vulnerable to challenge from anybody who thinks they have evidence to contradict a theory? Even though scientific knowledge is not sacred, the scientific process is. No observation is considered valid unless it can be reproduced by another scientist working independently under the same conditions. The peer-review process ensures that all results reported by a scientist undergo strict scrutiny by others working in the same field. Thus it is the integrity of the scientific process that keeps scientific knowledge, despite its openness to change, firmly grounded in objectivity and logic.

Skill 12.3   Determining an appropriate scientific hypothesis or investigative design for addressing a given problem

The design of experiments includes the planning of all steps of the information gathering activity. It is best to start any experiment with a clearly stated and understood set of goals and objectives. This will help lead the experimenter into defining the specific data that needs to be collected, how the data will be collected, and how the data will be analyzed after collection. Specifically, the experimenter should determine the number of observations needed, and over what period. The variables affecting the data collection should also be outlined and a determination of which ones will be held constant, which ones will be varied, and which ones may be out of the experimenter's control.

A scientific control improves the integrity of an experiment by isolating each variable in order to draw conclusions about the effect of the single variable in the result. As much as possible the experiments should be identical, except for the one variable being tested. If there are variables that are beyond the control of the experimenter, it is wise to identify them up front and attempt to mitigate their effect. Controls are generally one of two types, negative and positive. A negative control is used when a negative result is expected in an experiment. The negative control helps correlate a positive result with the variable being tested. For example, testing a drug on a group of rats and maintaining another group who are only given a placebo. A positive control is a sample that is known to produce a positive result to make sure the experiment is working as expected.

For example printing a page with he printers own drivers before testing the printer with another program.

Once the data collection process has been defined, collection of information can begin. It is wise to check the data periodically to ensure that the data is reasonable and is being collected appropriately. However, care must be taken that data that does not necessarily fit with expectations is not simply discarded. If the data does not fit expectations, corrections may be needed to the collection method or the expected results may not be accurate.

After collection, experimental data must be analyzed, interpreted and presented in a way that can be understood by others. Data analysis may include statistical methods, curve fitting, and dividing data into subsets. Data analysis transforms the information collected during experimentation with the goal of extracting useful information and drawing conclusions. Data interpretation is the method by which the data, in its raw and analyzed forms is reviewed for meaning and explanation. It is often necessary to look at historical information in the same area of study when interpreting the data from an experiment. Presenting the data is the final step in experimental design. In this way, information is related to others interested or affected by the results of the experiment. Graphical representations are useful to communicate information, although clear and concise language is always necessary to ensure the thorough understanding of your audience.

Theory
When a hypothesis survives many experimental tests to determine its validity, the hypothesis may evolve into a **theory**. A theory explains a body of facts and laws that are based on the facts. A theory also reliably predicts the outcome of related events in nature. For example, the law of conservation of matter and many other experimental observations led to a theory proposed early in the 19th century. This theory explained the conservation law by proposing that all matter is made up of atoms which are never created or destroyed in chemical reactions, only rearranged. This atomic theory also successfully predicted the behavior of matter in chemical reactions that had not been studied at the time. As a result, the atomic theory has stood for 200 years with only small modifications.

A theory also serves as a scientific **model**. A model can be a physical model made of wood or plastic, a computer program that simulates events in nature, or simply a mental picture of an idea. A model illustrates a theory and explains nature. For instance, in your science class you may develop a mental (and maybe a physical) model of the atom and its behavior. Outside of science, the word theory is often used to describe someone's unproven notion about something. In science, theory means much more. It is a thoroughly tested explanation of things and events observed in nature. A theory can never be proven true, but it can be proven untrue. All it takes to prove a theory untrue is to show an exception to the theory. The test of the hypothesis may be observations of phenomena or a model may be built to examine its behavior under certain circumstances.

Formulating problems

Although many discoveries happen by chance, the standard thought process of a scientist begins with forming a question to research. The more limited and clearly defined the question, the easier it is to set up an experiment to answer it. Scientific questions result from observations of events in nature or events observed in the laboratory. An **observation** is not just a look at what happens. It also includes measurements and careful records of the event. Records could include photos, drawings, or written descriptions. The observations and data collection lead to a question. In physics, observations almost always deal with the behavior of matter. Having arrived at a question, a scientist usually researches the scientific literature to see what is known about the question. Maybe the question has already been answered. The scientist then may want to test the answer found in the literature. Or, maybe the research will lead to a new question.

Sometimes the same observations are made over and over again and are always the same. For example, you can observe that daylight lasts longer in summer than in winter. This observation never varies. Such observations are called **laws** of nature. One of the most important scientific laws was discovered in the late 1700s. Chemists observed that no mass was ever lost or gained in chemical reactions. This law became known as the law of conservation of mass. Explaining this law was a major topic of scientific research in the early 19th century.

Forming a hypothesis

Once the question is formulated, take an educated guess about the answer to the problem or question. This 'best guess' is your hypothesis. A **hypothesis is a statement of a possible answer to the question**. It is a tentative explanation for a set of facts and can be tested by experiments. Although hypotheses are usually based on observations, they may also be based on a sudden idea or intuition.

Skill 12.4 Demonstrating knowledge of the principles and procedures for designing and carrying out scientific investigations

The scientific method is a logical set of steps that a scientist goes through to solve a problem. There are as many different scientific methods as there are scientists experimenting. However, there seems to be some pattern to their work. The scientific method is the process by which data is collected, interpreted and validated. While an inquiry may start at any point in this method and may not involve all of the steps here is the general pattern.

Experiment

An experiment tests the hypothesis to determine whether it may be a correct answer to the question or a solution to the problem. Some experiments may test the effect of one thing on another under controlled conditions. Such experiments have two variables. The experimenter controls one variable, called the **independent variable**. The other variable, the **dependent variable**, is the change caused by changing the independent variable. For example, suppose a researcher wanted to test the effect of vitamin A on the ability of rats to see in dim light. The independent variable would be the dose of Vitamin A added to the rats' diet. The dependent variable would be the intensity of light that causes the rats to react. All other factors, such as time, temperature, age, water given to the rats, the other nutrients given to the rats, and similar factors, are held constant. Scientists sometimes do short experiments "just to see what happens". Often, these are not formal experiments. Rather they are ways of making additional observations about the behavior of matter. A good test will try to manipulate as few variables as possible so as to see which variable is responsible for the result. This requires a second example of a **control**. A control is an extra setup in which all the conditions are the same except for the variable being tested.

In most experiments, scientists collect quantitative data, which is data that can be measured with instruments. They also collect qualitative data, descriptive information from observations other than measurements. Interpreting data and analyzing observations are important. If data is not organized in a logical manner, wrong conclusions can be drawn. Also, other scientists may not be able to follow your work or repeat your results.

Conclusion

Finally, a scientist must draw conclusions from the experiment. A conclusion must address the hypothesis on which the experiment was based. The conclusion states whether or not the data supports the hypothesis. If it does not, the conclusion should state what the experiment *did* show. If the hypothesis is not supported, the scientist uses the observations from the experiment to make a new or revised hypothesis. Then, new experiments are planned.

Skill 12.5   Recognizing the importance of and strategies for avoiding bias in scientific investigations

Experimental bias occurs when a researcher favors one particular outcome over another in an experimental setup. This can occur either purposefully or accidentally. Either way, it is important that experimental bias be avoided as much as possible in order to maintain the integrity of both the results of the experiment and the scientists themselves.

Types of experimental bias and how to avoid them

Systematic error is a type of bias that occurs accidentally. It results from imperfect equipment or technique that is consistent throughout the experiment. Systematic error decreases accuracy and causes a biased result that, on average, is too large or small. Potential bias, caused by systematic error, may be the result of poor instrument calibration or interference from other experimental factors. In contrast, **random error** occurs from limitations in equipment or techniques and usually results in a few imprecise data points in an experiment.

The **bandwagon effect** is a type of bias that can occur within a group of scientists. People working together on an experiment may make a poor decision that none would make individually. This can also occur when a scientist may censor him or herself in order to give an experimental result that is in line with the current accepted wisdom. In order to avoid this, each scientist must have the specific intent to report, objectively, the results of the experiment.

A **confirmation bias** is caused by a scientist "reading into" the results of an experiment in such a way that would confirm the hypothesis. This is also called the **observer-expectancy effect**. An opposite bias can occur when a scientist adjusts their hypothesis to suit their results, thus claiming their experiment a "success." This type of bias may also be avoided by a strong desire to maintain the integrity of the experiment.

Throughout the experiment, the scientist must be objective when reporting observations and results. Proper documentation is necessary to analyze experimental data. This will allow the experiment to be replicated by other scientists, thus improving the reliability of the original experiment.

Experimental controls

An **experimental control** is a common way to reduce bias in an experiment. It can prevent extraneous factors from impacting the outcome of the experiment. A **test sample** in a controlled experiment is the unknown that is compared against one or more **control samples**. Control samples should be selected to be as identical to the test sample as possible in every way other than the one variable being tested. A **negative control** is a control sample that is known to lack the effect. A **positive control** is known to contain the effect.

The experimental control helps to alleviate bias in an experiment by allowing the scientist to compare results between test and control samples. This will help the scientist to see the true results, and not just the desired results. However, control samples will not necessarily eliminate systematic bias, since that is commonly caused by an error that would be present in both the control and the test samples. For example, both samples would be inaccurate if the same scale is used to read results, but the scale is imbalanced.

TEACHER CERTIFICATION STUDY GUIDE

COMPETENCY 13.0 UNDERSTAND THE COLLECTION, ANALYSIS, AND COMMUNICATION OF SCIENTIFIC DATA

Skill 13.1 Identifying appropriate tools and units for measuring objects of substances

SI is an abbreviation of the French *Système International d'Unités* or the **International System of Units**. It is the most widely used system of units in the world and is the system used in science. The use of many SI units in the United States is increasing outside of science and technology. There are two types of SI units: **base units** and **derived units**. The base units are:

Quantity	Unit name	Symbol
Length	meter	m
Mass	kilogram	kg
Amount of substance	mole	mol
Time	second	s
Temperature	kelvin	K
Electric current	ampere	A
Luminous intensity	candela	cd

Decimal multiples of SI units are formed by attaching a **prefix** directly before the unit and a symbol prefix directly before the unit symbol. SI prefixes range from 10^{-24} to 10^{24}. Common prefixes you are likely to encounter in physics are shown below:

Factor	Prefix	Symbol	Factor	Prefix	Symbol
10^9	*giga—*	G	10^{-1}	*deci—*	d
10^6	*mega—*	M	10^{-2}	*centi—*	c
10^3	*kilo—*	k	10^{-3}	*milli—*	m
10^2	*hecto—*	h	10^{-6}	*micro—*	μ
10^1	*deca—*	da	10^{-9}	*nano—*	n
			10^{-12}	*pico—*	p

Example: 0.0000004355 meters is 4.355×10^{-7} m or 435.5×10^{-9} m. This length is also 435.5 nm or 435.5 nanometers.

Expressing volume in liters is helpful in cases like these. There is no power on the unit of liters, therefore: $1 L = 10^3$ mL $= 10^6$ μL $= 10^9$ nL.

Converting cubic meters to liters gives 8×10^{-12} m$^3 \times \dfrac{10^3 \text{ L}}{1 \text{ m}^3} = 8 \times 10^{-9}$ L.

The crystal's volume is 8 nanoliters (8 nL).

PHYSICS

Example: Determine the ideal gas constant, R, in L•atm/(mol•K) from its SI value of 8.3144 J/(mol•K).

Solution: One joule is identical to one m^3•Pa (see the table on the previous page).

$$8.3144 \frac{m^3 \cdot Pa}{mol \cdot K} \times \frac{1000 \text{ L}}{1 \text{ m}^3} \times \frac{1 \text{ atm}}{101325 \text{ Pa}} = 0.082057 \frac{L \cdot atm}{mol \cdot K}$$

The **order of magnitude** is a familiar concept in scientific estimation and comparison. It refers to a category of scale or size of an amount, where each category contains values of a fixed ratio to the categories before or after. The most common ratio is 10. Orders of magnitude are typically used to make estimations of a number. For example, if two numbers differ by one order of magnitude, one number is 10 times larger than the other. If they differ by two orders of magnitude the difference is 100 times larger or smaller, and so on. It follows that two numbers have the same order of magnitude if they differ by less than 10 times the size.

To estimate the order of magnitude of a physical quantity, you round the its value to the nearest power of 10. For example, in estimating the human population of the earth, you may not know if it is 5 billion or 12 billion, but a reasonable order of magnitude estimate is 10 billion. Similarly, you may know that Saturn is much larger than Earth and can estimate that it has approximately 100 times more mass, or that its mass is 2 orders of magnitude larger. The actual number is 95 times the mass of earth.

Below are the dimensions of some familiar objects expressed in orders of magnitude.

Physical Item	Size	Order of Magnitude (meters)
Diameter of a hydrogen atom	100 picometers	10^{-10}
Size of a bacteria	1 micrometer	10^{-6}
Size of a raindrop	1 millimeter	10^{-3}
Width of a human finger	1 centimeter	10^{-2}
Height of Washington Monument	100 meters	10^{2}
Height of Mount Everest	10 kilometers	10^{4}
Diameter of Earth	10 million meters	10^{7}
One light year	1 light year	10^{16}

The name "kilogram" occurs for the SI base unit of mass for historical reasons. Derived units are formed from the kilogram, but appropriate decimal prefixes are attached to the word "gram." Derived units measure a quantity that may be **expressed in terms of other units**. Some derived units important for physics are:

Derived quantity	Unit name	Expression in terms of other units	Symbol
Area	square meter	m^2	
Volume	cubic meter	m^3	
	liter	$dm^3 = 10^{-3} m^3$	L or l
Mass	unified atomic mass unit	$(6.022 \times 10^{23})^{-1}$ g	u or Da
Time	minute	60 s	min
	hour	60 min = 3600 s	h
	day	24 h = 86400 s	d
Speed	meter per second	m/s	
Acceleration	meter per second squared	m/s^2	
Temperature*	degree Celsius	K	°C
Mass density	gram per liter	$g/L = 1 kg/m^3$	
Force	newton	$m \cdot kg/s^2$	N
Pressure	pascal	$N/m^2 = kg/(m \cdot s^2)$	Pa
	standard atmosphere§	101325 Pa	atm
Energy, Work, Heat	joule	$N \cdot m = m^3 \cdot Pa = m^2 \cdot kg/s^2$	J
	nutritional calorie§	4184 J	Cal
Heat (molar)	joule per mole	J/mol	
Heat capacity, entropy	joule per kelvin	J/K	
Heat capacity (molar), entropy (molar)	joule per mole kelvin	J/(mol·K)	
Specific heat	joule per kilogram kelvin	J/(kg·K)	
Power	watt	J/s	W
Electric charge	coulomb	s·A	C
Electric potential, electromotive force	volt	W/A	V
Viscosity	pascal second	Pa·s	
Surface tension	newton per meter	N/m	

*Temperature differences in Kelvin are the same as those differences in degrees Celsius. To obtain degrees Celsius from Kelvin, subtract 273.15. Differentiate *m* and meters (m) by context.
§These are commonly used non-SI units.

PHYSICS

Skill 13.2 **Recognizing potential safety hazards and procedures for the safe and proper use of scientific tools, instruments, chemical, and other materials in investigations**

Safety is a learned behavior and must be incorporated into instructional plans. Measures of prevention and procedures for dealing with emergencies in hazardous situations have to be in place and readily available for reference. Copies of these must be given to all people concerned, such as administrators and students.

The single most important aspect of safety is planning and anticipating various possibilities and preparing for the eventuality. Any Physics teacher/educator planning on doing an experiment must try it before the students do it. In the event of an emergency, quick action can prevent many disasters. The teacher/educator must be willing to seek help at once without any hesitation because sometimes it may not be clear that the situation is hazardous and potentially dangerous.

There are a number of procedures to prevent and correct any hazardous situation. There are several safety aids available commercially such as posters, safety contracts, safety tests, safety citations, texts on safety in secondary classroom/laboratories, hand books on safety and a host of other equipment. Another important thing is to check the laboratory and classroom for safety and report it to the administrators before staring activities/experiments. It is important that teachers and educators follow these guidelines to protect the students and to avoid most of the hazards. They have a responsibility to protect themselves as well. **There should be not any compromises in issues of safety.**

All science labs should contain the following items of **safety equipment**.
-Fire blanket that is visible and accessible
-Ground Fault Circuit Interrupters (GCFI) within two feet of water supplies
-Signs designating room exits
-Emergency shower providing a continuous flow of water
-Emergency eye wash station that can be activated by the foot or forearm
-Eye protection for every student and a means of sanitizing equipment
-Emergency exhaust fans providing ventilation to the outside of the building
-Master cut-off switches for gas, electric and compressed air. Switches must have permanently attached handles. Cut-off switches must be clearly labeled.
-An ABC fire extinguisher
-Storage cabinets for flammable materials
-Chemical spill control kit
-Fume hood with a motor that is spark proof
-Protective laboratory aprons made of flame retardant material
-Signs that will alert potential hazardous conditions
-Labeled containers for broken glassware, flammables, corrosives, and waste.

Students should wear safety goggles when performing dissections, heating, or while using acids and bases. Hair should always be tied back and objects should never be placed in the mouth. Food should not be consumed while in the laboratory. Hands should always be washed before and after laboratory experiments. In case of an accident, eye washes and showers should be used for eye contamination or a chemical spill that covers the student's body. Small chemical spills should only be contained and cleaned by the teacher. Kitty litter or a chemical spill kit should be used to clean spill. For large spills, the school administration and the local fire department should be notified. Biological spills should only be handled by the teacher. Contamination with biological waste can be cleaned by using bleach when appropriate. Accidents and injuries should always be reported to the school administration and local health facilities. The severity of the accident or injury will determine the course of action to pursue.

All laboratory solutions should be prepared as directed in the lab manual. Care should be taken to avoid contamination. All glassware should be rinsed thoroughly with distilled water before using and cleaned well after use. All solutions should be made with distilled water as tap water contains dissolved particles that may affect the results of an experiment. Unused solutions should be disposed of according to local disposal procedures.

The "Right to Know Law" covers science teachers who work with potentially hazardous chemicals. Briefly, the law states that employees must be informed of potentially toxic chemicals. An inventory must be made available if requested. The inventory must contain information about the hazards and properties of the chemicals. This inventory is to be checked against the "Substance List". Training must be provided on the safe handling and interpretation of the **Material Safety Data Sheet (MSDS)**.

The following chemicals are potential carcinogens and not allowed in school facilities: Acrylonitriel, Arsenic compounds, Asbestos, Bensidine, Benzene, Cadmium compounds, Chloroform, Chromium compounds, Ethylene oxide, Ortho-toluidine, Nickel powder, and Mercury.

Chemicals should not be stored on bench tops or heat sources. They should be stored in groups based on their reactivity with one another and in protective storage cabinets. All containers within the lab must be labeled. Suspect and known carcinogens must be labeled as such and segregated within trays to contain leaks and spills.

Chemical waste should be disposed of in properly labeled containers. Waste should be separated based on their reactivity with other chemicals.

Biological material should never be stored near food or water used for human consumption. All biological material should be appropriately labeled. All blood and body fluids should be put in a well-contained container with a secure lid to prevent leaking. All biological waste should be disposed of in biological hazardous waste bags.

Material safety data sheets are available for every chemical and biological substance. These are available directly from the company of acquisition or the internet. The manuals for equipment used in the lab should be read and understood before using them.

Skill 13.3   Recognizing the concepts of precision, accuracy, and error and identifying potential sources of error in gathering and recording data

Precision is a measure of how similar repeated measurements from a given device or technique are. Note that this is distinguished from **accuracy** which refers to how close to "correct" a measuring device or technique is. Thus, accuracy can be tested by measuring a known quantity (a standard) and determining how close the value provided by the measuring device is. To determine precision, however, we must make multiple measurements of the same sample. The precision of an instrument is typically given in terms of its standard error or standard deviation. Precision is typically divided into reproducibility and repeatability. These concepts are subtly different and are defined as follows:

Repeatability: Variation observed in measurements made over a short period of time while trying to keep all conditions the same (including using the same instrument, the same environmental conditions, and the same operator)

Reproducibility: Variation observed in measurements taken over a long time period in a variety of different settings (different places and environments, using different instruments and operators)

Both repeatability and reproducibility can be estimated by taking multiple measurements under the conditions specified above. Using the obtained values, standard deviation can be calculated using the formula:

$$\sigma = \sqrt{\frac{1}{N}\sum_{i=1}^{N}(x_i - \overline{x})^2}$$

where σ = standard deviation
   N = the number of measurements
   x_i = the individual measured values
   $\overline{x}$ = the average value of the measured quantity

To obtain a reliable estimate of standard deviation, N, the number of samples, should be fairly large. We can use statistical methods to determine a confidence interval on our measurements. A typical confidence level for scientific investigations is 90% or 95%.

Often in scientific operations we want to determine a quantity that requires many steps to measure. Of course, each time we take a measurement there will be a certain associated error that is a function of the measuring device. Each of these errors contributes to an even greater one in the final value. This phenomenon is known as **propagation of error** or propagation of uncertainty.

A measured value is typically expressed in the form x±Δx, where Δx is the uncertainty or margin of error. What this means is that the value of the measured quantity lies somewhere between x-Δx and x+Δx, but our measurement techniques do not allow us any more precision. If several measurements are required to ultimately decide a value, we must use formulas to determine the total uncertainty that results from all the measurement errors. A few of these formulas for simple functions are listed below:

Formula	Uncertainty
$X = A \pm B$	$(\Delta X)^2 = (\Delta A)^2 + (\Delta B)^2$
$X = cA$	$\Delta X = c\Delta A$
$X = c(A \cdot B)$	$\left(\frac{\Delta X}{X}\right)^2 = \left(\frac{\Delta A}{A}\right)^2 + \left(\frac{\Delta B}{B}\right)^2$
$X = c\left(\frac{A}{B}\right)$	$\left(\frac{\Delta X}{X}\right)^2 = \left(\frac{\Delta A}{A}\right)^2 + \left(\frac{\Delta B}{B}\right)^2$

For example, if we wanted to determine the density of a small piece of metal we would have to measure its weight on a scale and then determine its volume by measuring the amount of water it displaces in a graduated cylinder. There will be error associated with measurements made by both the scale and the graduated cylinder. Let's suppose we took the following measurements:

Mass: 57± 0.5 grams
Volume: 23 ± 3 mm³

Since density is simply mass divided by the volume, we can determine its value to be:

$$\rho = \frac{m}{V} = \frac{57g}{23mm^3} = 2.5\frac{g}{mm^3}$$

Now we must calculate the uncertainty on this measurement, using the formula above:

$$\left(\frac{\Delta x}{x}\right)^2 = \left(\frac{\Delta A}{A}\right)^2 + \left(\frac{\Delta B}{B}\right)^2$$

$$\Delta x = \left(\sqrt{\left(\frac{\Delta A}{A}\right)^2 + \left(\frac{\Delta B}{B}\right)^2}\right) x = \left(\sqrt{\left(\frac{0.5g}{57g}\right)^2 + \left(\frac{3mm^3}{23mm^3}\right)^2}\right) \times 2.5\frac{g}{mm^3} = 0.3\frac{g}{mm^3}$$

Thus, the final value for the density of this object is 2.5 ± 0.3 g/mm³.

Skill 13.4 Identifying methods and criteria for organizing and analyzing data

Once the data collection process has been defined, collection of information can begin. It is wise to check the data periodically to ensure that the data is reasonable and is being collected appropriately. However, care must be taken that data that does not necessarily fit with expectations is not simply discarded. If the data does not fit expectations, corrections may be needed to the collection method or the expected results may not be accurate.

Advances in information technology, including software, have been immensely helpful in all the physical sciences. Information technology has allowed for the rapid communication of experimental results and hypotheses between scientists and engineers working in distant locations. For example, the ability to communicate rapidly by email and send digital images and other files means scientists can easily confer on new data and interpretations. They can easily find and consult with experts outside their own field who might have particular knowledge. Information technology has also made it easier to find already published experimental results; many journals have online editions and search engines make finding information on a given topic straightforward.

Specific software packages can also assist the advancement of science. Some important software and their appropriate uses are detailed below:

Spreadsheets

Spreadsheets contain multiple pages of grids into which numbers and formulas can be entered. They are excellent tools for storing data and doing some simple mathematical manipulations. Many spreadsheets now include graphical and statistical packages (see below), though often they are not as powerful as "stand alone" software for those uses. Spreadsheets are not typically used for complex statistical analysis or simulations, though they may have some rudimentary capabilities. The data would then be imported it to another, more specific program if further analysis were needed.

Databases

Like spreadsheets, databases are designed for the storage and manipulation of data. However, they are much more sophisticated and have increased security and accountability control. Though they may sometimes be useful for managing extremely large or shared datasets, databases are not typically useful to scientists.

Graphing packages

Graphing packages allow data to be transformed into a variety of visual representation. Because graphing is so important for recognizing trends and communicating experimental results, scientists make extensive use of this type of software. Most graphical packages will also perform basic modeling or statistical analysis (for instance, fitting a line or determining standard deviation). In some simple cases, this is appropriate, but most often, statistical packages are better suited for this type of analysis. Graphing software should usually be limited to displaying the results of this analysis (for example, with error bars).

Statistical packages

Statistical packages perform all types of statistical analysis to determine if significant differences exist between datasets. Thus, this type of software is extremely helpful to scientists especially when one is trying to analyze trends or determine the effects of a given factor in a large, complex experiment. Though this software can simplify the performance of a statistical test, the scientist must still carefully determine *which* statistical test will provide her with the answer she seeks. Though statistical software packages may have tools for data storage and graphing, it is usually preferable to use these only temporarily or for a "first-pass" and ultimately use a more specific software package for these tasks.

Simulators

Simulators assist with the creation of complex mathematical models of the natural world. They also allow for simulated experiments, in which the value of variables in the model are changed and hypothetical outcomes obtained. These simulators can be a tremendous boon to scientists when performing actual experiments would be expensive, unethical, or impossible. Though much important information can be obtained from these simulators, it is important that there findings be supported by experimental evidence whenever possible.

Additionally, many of the graphical representations discussed in **Skill 13.5** are useful for analyzing scientific data.

Skill 13.5 Identifying appropriate methods for communicating the outcomes of scientific investigations

In scientific investigations, it is often necessary to gather and analyze large data sets. We may need to manage data taken over long periods of time and under various different conditions. Appropriately organizing this data is necessary to identify trends and present the information to others. The uses and advantages of various graphic representations are discussed below.

Tables
Tables are excellent for organize data as it is being recorded and for storing data that need to be analyzed. In fact, almost all experimental data is initially organized into a table, such as in a lab notebook. Often, it is then entered into tables within spread sheets for further processing. Tables can be used for presenting data to others if the data set is fairly small or has been summarized (for example, presenting average values). However, for larger data sets, tabular presentation may be overwhelming. Finally, tables are not particularly useful for recognizing trends in data or for making them apparent to others.

Charts and Graphs
Charts and graphs are the best way to demonstrating trends or differences between groups. There are also useful for summarizing data and presenting it. In most types of graphs, it is also simple to indicate uncertainty of experimental data using error bars. Many types of charts and graphs are available to meet different needs. Three of the most common are scatterplots, bar charts, and pie charts. An example of each is shown.

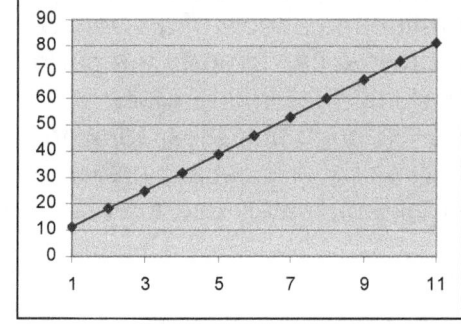

Scatterplots are typically shown on a Cartesian plane and are useful for demonstrating the relationship between two variables. A line chart (shown) is a special type of 2 dimensional scatter plot in which the data points are connected with a line to make a trend more apparent.

Linear regression is a common technique used by graphing software to translate tabular data into plots in which continuous lines are displayed even though data is available only at some discrete points. Regression is essentially a filling in of the gaps between data points by making a reasonable estimation of what the in-between values are using a standard mathematical process. In other words, it is finding the "best-fit" curve to represent experimental data. The graphs displayed below show how a straight line or a curve can be fitted to a set of discrete data points. There are many different regression algorithms that may be used and they differ from tool to tool. Some tools allow the user to decide what regression method will be used and what kind of curve (straight line, exponential etc.) will be used to fit the data.

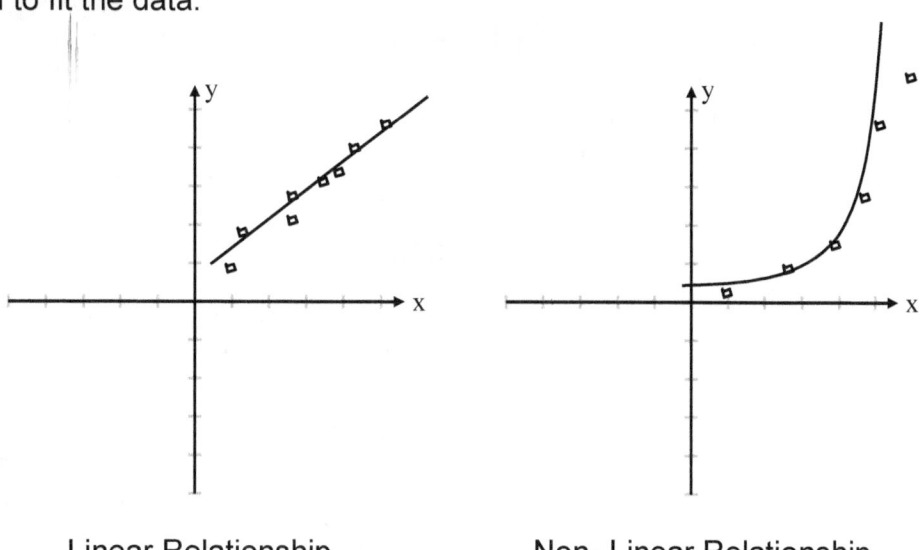

Linear Relationship   Non- Linear Relationship

Contrast the preceding graphs to the graph of a data set that shows no relationship between variables.

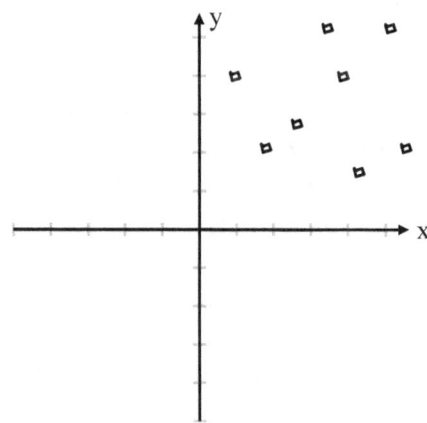

Bar charts can sometimes fill the same role as scatterplots, but are better suited to show values across different categories or different experimental conditions (especially where those conditions are described qualitatively, rather than quantified). Note the use of error bars in this example.

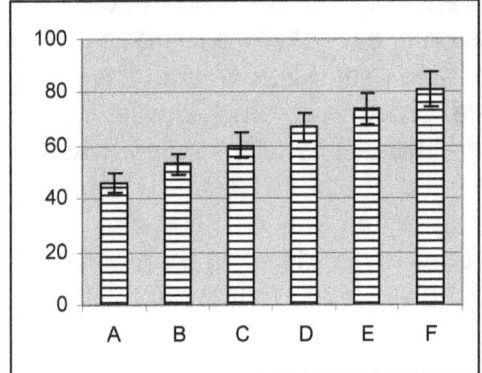

Finally, a pie chart is best used to present relative magnitudes or frequencies of several different conditions or events. They are most commonly used to show how various categories contribute to a whole.

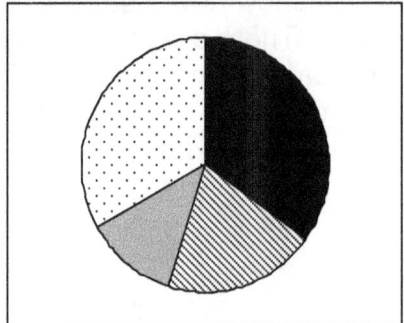

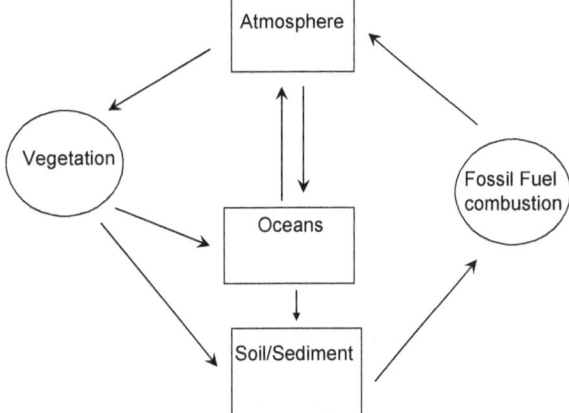

Diagrams

Diagrams not typically used to presenting the specifics of data. However, they are very good for demonstrating phenomena qualitatively. Diagrams make it easy to visualize the connections and relationships between various elements. They may also be used to demonstrate temporal relationships. For example, diagrams can be used to illustrate the operation of an internal combustion engine or the complex biochemical pathways of an enzyme's action. The diagram shown is a simplified version of the carbon cycle.

Models

While it is not specifically a type of graphical representation, developing a scientific model is the ultimate goal of all systematic investigations. A scientific model is a set of ideas that describes a natural process and is developed by empirical or theoretical methods. Models help scientists focus on the basic fundamental processes. They may be physical representations, such as a space-filling model of a molecule or a map, or they may be mathematical algorithms. Whatever form they take, scientific models are based on what is known about the systems or objects at the time that the models are constructed. Models usually evolve and are improved as scientific advances are made. Sometimes a model must be discarded because new findings show it to be misleading or incorrect.

Models are developed in an effort to explain how things work in nature. Because models are not the "real thing", they can never correctly represent the system or object in all respects. The amount of detail that they contain depends upon how the model will be used as well as the sophistication and skill of the scientist doing the modeling. If a model has too many details left out, its usefulness may be limited. But too many details may make a model too complicated to be useful.

Skill 13.6 Demonstrating familiarity with effective resources and strategies for reading to gain information about science-related topics and developing subject-area vocabulary

The type of resource that is appropriate for learning about a particular science-related topic depends, in large measure, on the level of understanding required. Additionally, an important factor is the relative newness or novelty of the topic of interest.

Popular articles and books

When investigating an unfamiliar science-related topic, articles in popular science magazines, such as *Science*, *Popular Science* and even *National Geographic*, can be a helpful starting point. Such magazines often provide articles on scientific discoveries and ideas that cover the topic at an introductory level that is accessible to most readers. In addition to presenting some of the basic (and, sometimes, more advanced) concepts, these articles almost invariably also use at least some vocabulary associated with the topic. Many popular books also serve the same purpose as their counterparts in periodical literature, but provide a more thorough presentation of the topic. Although such popular articles and books may provide a good starting point when learning about some subject, they seldom offer a solid foundational or mathematical understanding. As a result, other sources must be consulted for more in-depth knowledge.

Textbooks

Student-oriented textbooks are a good resource for acquiring a more fundamental overview of a topic related to science. Textbooks regularly define critical terms and vocabulary, and very often include a bibliography that can be used as a starting point for more in-depth reading or research. Although textbooks often provide a reasonable theoretical foundation for understanding a particular subject, the discussion of relevant examples, current issues, and other specific information is often highly limited for the sake of brevity. Additionally, even classic textbooks that adequately cover certain topics may be out of date with respect to recent advances and controversies. As a result, textbooks are not an ideal source of information for highly specific or novel research. Some books that cover more specific topics are available and may be better resources than textbooks, however.

Peer-reviewed literature

The most up-to-date and informative literature dealing with specific issues and controversies in science is found in the peer-reviewed publications. These publications report new discoveries and findings and provide the most unadulterated view of the current status of a particular field of science. Due to the high-level technical nature of these publications, however, they are often inaccessible to readers not already familiar with at least the basics of the field. Also, knowledge of most of the technical vocabulary is assumed in journal articles. Nevertheless, with appropriate research, most papers can be unraveled and information can be gleaned from their pages. Furthermore, many of these technical articles have extensive bibliographies that can be used to do further background or related research.

Strategies for acquiring information

How one goes about studying or learning about a specific science-related topic or field depends largely on the specificity and technical level of the desired understanding, as well as on the background of the investigator. Someone new to a particular field may need to start at the level of a popular article or book and then build up, by way of textbooks and other middle-level resources, to peer-reviewed literature. Those with a solid background in a field who are seeking to do original research on a specific topic may find it sufficient to go directly to the peer-reviewed literature. The best strategy for learning about a science-related topic must take into account these and other considerations.

Sample Test

DIRECTIONS: Read each item and select the best response.

1. **Which statement best describes a valid approach to testing a scientific hypothesis?**
   (Easy)

   A. Use computer simulations to verify the hypothesis

   B. Perform a mathematical analysis of the hypothesis

   C. Design experiments to test the hypothesis

   D. All of the above

2. **Which description best describes the role of a scientific model of a physical phenomenon?**
   (Average Rigor)

   A. An explanation that provides a reasonably accurate approximation of the phenomenon

   B. A theoretical explanation that describes exactly what is taking place

   C. A purely mathematical formulation of the phenomenon

   D. A predictive tool that has no interest in what is actually occurring

3. **Which situation calls might best be described as involving an ethical dilemma for a scientist?**
   (Rigorous)

   A. Submission to a peer-review journal of a paper that refutes an established theory

   B. Synthesis of a new radioactive isotope of an element

   C. Use of a computer for modeling a newly-constructed nuclear reactor

   D. Use of a pen-and-paper approach to a difficult problem

4. **Which of the following is not a key purpose for the use of open communication about and peer-review of the results of scientific investigations?**
(Average Rigor)

   A. Testing, by other scientists, of the results of an investigation for the purpose of refuting any evidence contrary to an established theory

   B. Testing, by other scientists, of the results of an investigation for the purpose of finding or eliminating any errors in reasoning or measurement

   C. Maintaining an open, public process to better promote honesty and integrity in science

   D. Provide a forum to help promote progress through mutual sharing and review of the results of investigations

5. **Which of the following aspects of the use of computers for collecting experimental data is not a concern for the scientist?**
(Rigorous)

   A. The relative speeds of the processor, peripheral, memory storage unit and any other components included in data acquisition equipment

   B. The financial cost of the equipment, utilities and maintenance

   C. Numerical error due to a lack of infinite precision in digital equipment

   D. The order of complexity of data analysis algorithms

6. **If a particular experimental observation contradicts a theory, what is the most appropriate approach that a physicist should take?**
   (Average Rigor)

   A. Immediately reject the theory and begin developing a new theory that better fits the observed results

   B. Report the experimental result in the literature without further ado

   C. Repeat the observations and check the experimental apparatus for any potential faulty components or human error, and then compare the results once more with the theory

   D. Immediately reject the observation as in error due to its conflict with theory

7. **Which of the following is *not* an SI unit?**
   (Average Rigor)

   A. Joule

   B. Coulomb

   C. Newton

   D. Erg

8. **Which of the following best describes the relationship of precision and accuracy in scientific measurements?**
   (Easy)

   A. Accuracy is how well a particular measurement agrees with the value of the actual parameter being measured; precision is how well a particular measurement agrees with the average of other measurements taken for the same value

   B. Precision is how well a particular measurement agrees with the value of the actual parameter being measured; accuracy is how well a particular measurement agrees with the average of other measurements taken for the same value

   C. Accuracy is the same as precision

   D. Accuracy is a measure of numerical error; precision is a measure of human error

9. Which statement best describes a rationale for the use of statistical analysis in characterizing the numerical results of a scientific experiment or investigation?
(Average Rigor)

   A. Experimental results need to be adjusted, through the use of statistics, to conform to theoretical predictions and computer models

   B. Since experiments are prone to a number of errors and uncertainties, statistical analysis provides a method for characterizing experimental measurements by accounting for or quantifying these undesirable effects

   C. Experiments are not able to provide any useful information, and statistical analysis is needed to impose a theoretical framework on the results

   D. Statistical analysis is needed to relate experimental measurements to computer-simulated values

10. Which statement best characterizes the relationship of mathematics and experimentation in physics?
(Easy)

   A. Experimentation has no bearing on the mathematical models that are developed for physical phenomena

   B. Mathematics is a tool that assists in the development of models for various physical phenomena as they are studied experimentally, with observations of the phenomena being a test of the validity of the mathematical model

   C. Mathematics is used to test the validity of experimental apparatus for physical measurements

   D. Mathematics is an abstract field with no relationship to concrete experimentation

TEACHER CERTIFICATION STUDY GUIDE

11. Which of the following mathematical tools would not typically be used for the analysis of an electromagnetic phenomenon?
    (Rigorous)

    A. Trigonometry

    B. Vector calculus

    C. Group theory

    D. Numerical methods

12. For a problem that involves parameters that vary in rate with direction and location, which of the following mathematical tools would most likely be of greatest value?
    (Rigorous)

    A. Trigonometry

    B. Numerical analysis

    C. Group theory

    D. Vector calculus

13. Which of the following devices would be best suited for an experiment designed to measure alpha particle emissions from a sample?
    (Average Rigor)

    A. Photomultiplier tube

    B. Thermocouple

    C. Geiger-Müller tube

    D. Transistor

14. Which of the following experiments presents the most likely cause for concern about laboratory safety?
    (Average Rigor)

    A. Computer simulation of a nuclear reactor

    B. Vibration measurement with a laser

    C. Measurement of fluorescent light intensity with a battery-powered photodiode circuit

    D. Ambient indoor ionizing radiation measurement with a Geiger counter.

PHYSICS

15. A brick and hammer fall from a ledge at the same time. They would be expected to:
    (Easy)

    A. Reach the ground at the same time

    B. Accelerate at different rates due to difference in weight

    C. Accelerate at different rates due to difference in potential energy

    D. Accelerate at different rates due to difference in kinetic energy

16. A baseball is thrown with an initial velocity of 30 m/s at an angle of 45°. Neglecting air resistance, how far away will the ball land?
    (Rigorous)

    A. 92 m

    B. 78 m

    C. 65 m

    D. 46 m

17. A skateboarder accelerates down a ramp, with constant acceleration of two meters per second squared, from rest. The distance in meters, covered after four seconds, is:
    (Rigorous)

    A. 10

    B. 16

    C. 23

    D. 37

18. When acceleration is plotted versus time, the area under the graph represents:
    (Average Rigor)

    A. Time

    B. Distance

    C. Velocity

    D. Acceleration

19. An inclined plane is tilted by gradually increasing the angle of elevation θ, until the block will slide down at a constant velocity. The coefficient of friction, μ_k, is given by:
    (Rigorous)

    A. cos θ

    B. sin θ

    C. cosecant θ

    D. tangent θ

20. An object traveling through air loses part of its energy of motion due to friction. Which statement best describes what has happened to this energy?
    (Easy)

    A. The energy is destroyed

    B. The energy is converted to static charge

    C. The energy is radiated as electromagnetic waves

    D. The energy is lost to heating of the air

21. The weight of an object on the earth's surface is designated x. When it is two earth's radii from the surface of the earth, its weight will be:
    (Rigorous)

    A. x/4

    B. x/9

    C. 4x

    D. 16x

22. Which of the following units is not used to measure torque?
    (Average Rigor)

    A. slug ft

    B. lb ft

    C. N m

    D. dyne cm

23. A uniform pole weighing 100 grams, that is one meter in length, is supported by a pivot at 40 centimeters from the left end. In order to maintain static position, a 200 gram mass must be placed _____ centimeters from the left end.
(Rigorous)

   A. 10

   B. 45

   C. 35

   D. 50

24. The magnitude of a force is:
(Easy)

   A. Directly proportional to mass and inversely to acceleration

   B. Inversely proportional to mass and directly to acceleration

   C. Directly proportional to both mass and acceleration

   D. Inversely proportional to both mass and acceleration

25. A projectile with a mass of 1.0 kg has a muzzle velocity of 1500.0 m/s when it is fired from a cannon with a mass of 500.0 kg. If the cannon slides on a frictionless track, it will recoil with a velocity of ____ m/s.
(Rigorous)

   A. 2.4

   B. 3.0

   C. 3.5

   D. 1500

26. A car (mass m_1) is driving at velocity v, when it smashes into an unmoving car (mass m_2), locking bumpers. Both cars move together at the same velocity. The common velocity will be given by:
(Rigorous)

   A. $m_1 v / m_2$

   B. $m_2 v / m_1$

   C. $m_1 v / (m_1 + m_2)$

   D. $(m_1 + m_2) v / m_1$

27. A satellite is in a circular orbit above the earth. Which statement is false?
(Average Rigor)

   A. An external force causes the satellite to maintain orbit.

   B. The satellite's inertia causes it to maintain orbit.

   C. The satellite is accelerating toward the earth.

   D. The satellite's velocity and acceleration are not in the same direction.

28. A 100 g mass revolving around a fixed point, on the end of a 0.5 meter string, circles once every 0.25 seconds. What is the magnitude of the centripetal acceleration?
(Average Rigor)

   A. 1.23 m/s^2

   B. 31.6 m/s^2

   C. 100 m/s^2

   D. 316 m/s^2

29. Which statement best describes the relationship of simple harmonic motion to a simple pendulum of length L, mass m and displacement of arc length s?
(Average Rigor)

   A. A simple pendulum cannot be modeled using simple harmonic motion

   B. A simple pendulum may be modeled using the same expression as Hooke's law for displacement s, but with a spring constant equal to the tension on the string

   C. A simple pendulum may be modeled using the same expression as Hooke's law but with a spring constant equal to m g/L

   D. A simple pendulum typically does not undergo simple harmonic motion

30. A mass of 2 kg connected to a spring undergoes simple harmonic motion at a frequency of 3 Hz. What is the spring constant?
(Average Rigor)

   A. 6 kg/s^2

   B. 18 kg/s^2

   C. 710 kg/s^2

   D. 1000 kg/s^2

31. The kinetic energy of an object is _____ proportional to its _____.
(Average Rigor)

A. Inversely...inertia

B. Inversely...velocity

C. Directly...mass

D. Directly...time

32. A force is given by the vector 5 N x + 3 N y (where x and y are the unit vectors for the x- and y- axes, respectively). This force is applied to move a 10 kg object 5 m, in the x direction. How much work was done?
(Rigorous)

A. 250 J

B. 400 J

C. 40 J

D. 25 J

33. An office building entry ramp uses the principle of which simple machine?
(Easy)

A. Lever

B. Pulley

C. Wedge

D. Inclined Plane

34. If the internal energy of a system remains constant, how much work is done by the system if 1 kJ of heat energy is added?
(Average Rigor)

A. 0 kJ

B. -1 kJ

C. 1 kJ

D. 3.14 kJ

TEACHER CERTIFICATION STUDY GUIDE

35. A calorie is the amount of heat energy that will:
    (Easy)

    A. Raise the temperature of one gram of water from 14.5° C to 15.5° C.

    B. Lower the temperature of one gram of water from 16.5° C to 15.5° C

    C. Raise the temperature of one gram of water from 32° F to 33° F

    D. Cause water to boil at two atmospheres of pressure.

36. An ice block at 0° Celsius is dropped into 100 g of liquid water at 18° Celsius. When thermal equilibrium is achieved, only liquid water at 0° Celsius is left. What was the mass, in grams, of the original block of ice?
    Given:
    1. Heat of fusion of ice = 80 cal/g
    2. Heat of vaporization of ice = 540 cal/g
    3. Specific Heat of ice = 0.50 cal/g°C
    4. Specific Heat of water = 1 cal/g°C

    (Rigorous)

    A. 2.0

    B. 5.0

    C. 10.0

    D. 22.5

37. Heat transfer by electromagnetic waves is termed:
    (Easy)

    A. Conduction

    B. Convection

    C. Radiation

    D. Phase Change

38. A cooking thermometer in an oven works because the metals it is composed of have different:
    (Average Rigor)

    A. Melting points

    B. Heat convection

    C. Magnetic fields

    D. Coefficients of expansion

PHYSICS

39. Which of the following is not an assumption upon which the kinetic-molecular theory of gases is based?
(Rigorous)

   A. Quantum mechanical effects may be neglected

   B. The particles of a gas may be treated statistically

   C. The particles of the gas are treated as very small masses

   D. Collisions between gas particles and container walls are inelastic

40. What is temperature?
(Average Rigor)

   A. Temperature is a measure of the conductivity of the atoms or molecules in a material

   B. Temperature is a measure of the kinetic energy of the atoms or molecules in a material

   C. Temperature is a measure of the relativistic mass of the atoms or molecules in a material

   D. Temperature is a measure of the angular momentum of electrons in a material

41. Solids expand when heated because:
(Rigorous)

   A. Molecular motion causes expansion

   B. PV = nRT

   C. Magnetic forces stretch the chemical bonds

   D. All material is effectively fluid

42. What should be the behavior of an electroscope, which has been grounded in the presence of a positively charged object (1), after the ground connection is removed and then the charged object is removed from the vicinity (2)? *(Average Rigor)*

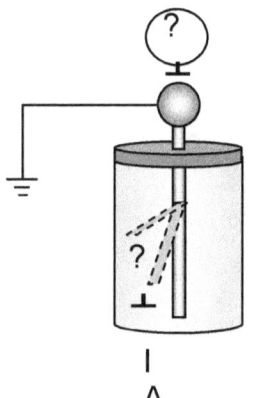
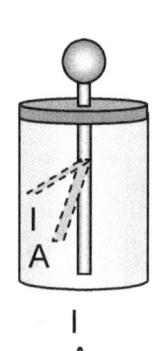

A. The metal leaf will start deflected (1) and then relax to an undeflected position (2)

B. The metal leaf will start in an undeflected position (1) and then be deflected (2)

C. The metal leaf will remain undeflected in both cases

D. The metal leaf will be deflected in both cases

43. The electric force in Newtons, on two small objects (each charged to – 10 microCoulombs and separated by 2 meters) is: *(Rigorous)*

A. 1.0

B. 9.81

C. 31.0

D. 0.225

44. A 10 ohm resistor and a 50 ohm resistor are connected in parallel. If the current in the 10 ohm resistor is 5 amperes, the current (in amperes) running through the 50 ohm resistor is: *(Rigorous)*

A. 1

B. 50

C. 25

D. 60

45. How much power is dissipated through the following resistive circuit?
    (Average Rigor)

    A. 0 W

    B. 0.22 W

    C. 0.31 W

    D. 0.49 W

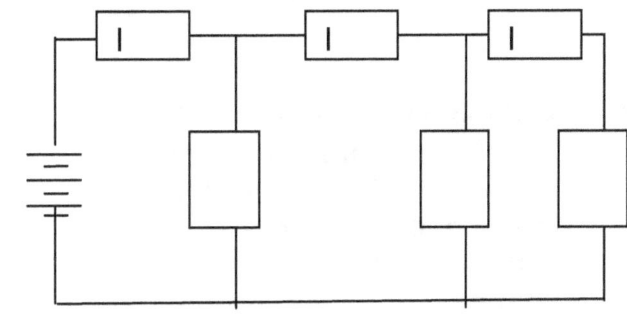

46. The greatest number of 100 watt lamps that can be connected in parallel with a 120 volt system without blowing a 5 amp fuse is:
    (Rigorous)

    A. 24

    B. 12

    C. 6

    D. 1

47. Which of the following statements may be taken as a legitimate inference based upon the Maxwell equation that states $\nabla \cdot \mathbf{B} = 0$?
    (Average Rigor)

    A. The electric and magnetic fields are decoupled

    B. The electric and magnetic fields are mediated by the W boson

    C. There are no photons

    D. There are no magnetic monopoles

48. What effect might an applied external magnetic field have on the magnetic domains of a ferromagnetic material?
    (Rigorous)

    A. The domains that are not aligned with the external field increase in size, but those that are aligned decrease in size

    B. The domains that are not aligned with the external field decrease in size, but those that are aligned increase in size

    C. The domains align perpendicular to the external field

    D. There is no effect on the magnetic domains

49. What is the effect of running current in the same direction along two parallel wires, as shown below?
(Rigorous)

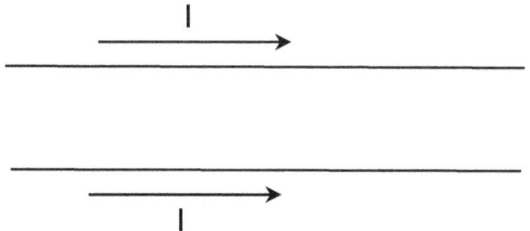

A. There is no effect

B. The wires attract one another

C. The wires repel one another

D. A torque is applied to both wires

50. The current induced in a coil is defined by which of the following laws?
(Easy)

A. Lenz's Law

B. Burke's Law

C. The Law of Spontaneous Combustion

D. Snell's Law

51. A light bulb is connected in series with a rotating coil within a magnetic field. The brightness of the light may be increased by any of the following except:
(Average Rigor)

A. Rotating the coil more rapidly.

B. Using more loops in the coil.

C. Using a different color wire for the coil.

D. Using a stronger magnetic field.

52. What is the direction of the magnetic field at the center of the loop of current (I) shown below (i.e., at point A)?
(Easy)

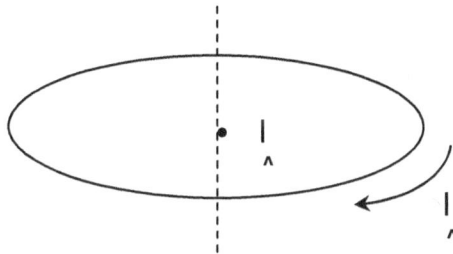

A. Down, along the axis (dotted line)

B. Up, along the axis (dotted line)

C. The magnetic field is oriented in a radial direction

D. There is no magnetic field at point A

53. The use of two circuits next to each other, with a change in current in the primary circuit, demonstrates:
    (Rigorous)

    A. Mutual current induction

    B. Dielectric constancy

    C. Harmonic resonance

    D. Resistance variation

54. A semi-conductor allows current to flow:
    (Easy)

    A. Never

    B. Always

    C. As long as it stays below a maximum temperature

    D. When a minimum voltage is applied

55. All of the following use semi-conductor technology, except a(n):
    (Average Rigor)

    A. Transistor

    B. Diode

    C. Capacitor

    D. Operational Amplifier

56. A wave generator is used to create a succession of waves. The rate of wave generation is one every 0.33 seconds. The period of these waves is:
    (Average Rigor)

    A. 2.0 seconds

    B. 1.0 seconds

    C. 0.33 seconds

    D. 3.0 seconds

57. An electromagnetic wave propagates through a vacuum. Independent of its wavelength, it will move with constant:
(Easy)

   A. Acceleration

   B. Velocity

   C. Induction

   D. Sound

58. A wave has speed 60 m/s and wavelength 30,000 m. What is the frequency of the wave?
(Average Rigor)

   A. 2.0×10^{-3} Hz

   B. 60 Hz

   C. 5.0×10^2 Hz

   D. 1.8×10^6 Hz

59. Rainbows are created by:
(Easy)

   A. Reflection, dispersion, and recombination

   B. Reflection, resistance, and expansion

   C. Reflection, compression, and specific heat

   D. Reflection, refraction, and dispersion

60. Which of the following is *not* a legitimate explanation for refraction of light rays at boundaries between different media?
(Rigorous)

   A. Light seeks the path of least time between two different points

   B. Due to phase matching and other boundary conditions, plane waves travel in different directions on either side of the boundary, depending on the material parameters

   C. The electric and magnetic fields become decoupled at the boundary

   D. Light rays obey Snell's law

61. A stationary sound source produces a wave of frequency F. An observer at position A is moving toward the horn, while an observer at position B is moving away from the horn. Which of the following is true?
(Rigorous)

   A. $F_A < F < F_B$

   B. $F_B < F < F_A$

   C. $F < F_A < F_B$

   D. $F_B < F_A < F$

62. A monochromatic ray of light passes from air to a thick slab of glass (n = 1.41) at an angle of 45° from the normal. At what angle does it leave the air/glass interface?
(Rigorous)

   A. 45°

   B. 30°

   C. 15°

   D. 55°

63. If one sound is ten decibels louder than another, the ratio of the intensity of the first to the second is:
(Average Rigor)

   A. 20:1

   B. 10:1

   C. 1:1

   D. 1:10

64. The velocity of sound is greatest in:
(Average Rigor)

   A. Water

   B. Steel

   C. Alcohol

   D. Air

65. A vibrating string's frequency is _____ proportional to the _____.
    (Rigorous)

    A. Directly; Square root of the tension

    B. Inversely; Length of the string

    C. Inversely; Squared length of the string

    D. Inversely; Force of the plectrum

66. Which of the following apparatus can be used to measure the wavelength of a sound produced by a tuning fork?
    (Average Rigor)

    A. A glass cylinder, some water, and iron filings

    B. A glass cylinder, a meter stick, and some water

    C. A metronome and some ice water

    D. A comb and some tissue

67. The highest energy is associated with:
    (Easy)

    A. UV radiation

    B. Yellow light

    C. Infrared radiation

    D. Gamma radiation

68. An object two meters tall is speeding toward a plane mirror at 10 m/s. What happens to the image as it nears the surface of the mirror?
    (Rigorous)

    A. It becomes inverted.

    B. The Doppler Effect must be considered.

    C. It remains two meters tall.

    D. It changes from a real image to a virtual image.

69. Automobile mirrors that have a sign, "objects are closer than they appear" say so because:
    (Rigorous)

    A. The real image of an obstacle, through a converging lens, appears farther away than the object.

    B. The real or virtual image of an obstacle, through a converging mirror, appears farther away than the object.

    C. The real image of an obstacle, through a diverging lens, appears farther away than the object.

    D. The virtual image of an obstacle, through a diverging mirror, appears farther away than the object.

70. If an object is 20 cm from a convex lens whose focal length is 10 cm, the image is:
    (Rigorous)

    A. Virtual and upright

    B. Real and inverted

    C. Larger than the object

    D. Smaller than the object

71. The constant of proportionality between the energy and the frequency of electromagnetic radiation is known as the:
    (Easy)

    A. Rydberg constant

    B. Energy constant

    C. Planck constant

    D. Einstein constant

72. Which phenomenon was first explained using the concept of quantization of energy, thus providing one of the key foundational principles for the later development of quantum theory?
    (Rigorous)

    A. The photoelectric effect

    B. Time dilation

    C. Blackbody radiation

    D. Magnetism

73. Which statement best describes why population inversion is necessary for a laser to operate?
    (Rigorous)

    A. Population inversion prevents too many electrons from being excited into higher energy levels, thus preventing damage to the gain medium.

    B. Population inversion maintains a sufficient number of electrons in a higher energy state so as to allow a significant amount of stimulated emission.

    C. Population inversion prevents the laser from producing coherent light.

    D. Population inversion is not necessary for the operation of most lasers.

74. Bohr's theory of the atom was the first to quantize:
    (Average Rigor)

    A. Work
    B. Angular Momentum
    C. Torque
    D. Duality

75. Two neutral isotopes of a chemical element have the same numbers of:
    (Easy)

    A. Electrons and Neutrons
    B. Electrons and Protons
    C. Protons and Neutrons
    D. Electrons, Neutrons, and Protons

76. When a radioactive material emits an alpha particle only, its atomic number will:
    (Average Rigor)

    A. Decrease
    B. Increase
    C. Remain unchanged
    D. Change randomly

77. Ten grams of a sample of a radioactive material (half-life = 12 days) were stored for 48 days and re-weighed. The new mass of material was:
    (Rigorous)

    A. 1.25 g
    B. 2.5 g
    C. 0.83 g
    D. 0.625 g

78. **Which of the following pairs of elements are not found to fuse in the centers of stars?**
    (Average Rigor)

    A. Oxygen and Helium

    B. Carbon and Hydrogen

    C. Beryllium and Helium

    D. Cobalt and Hydrogen

79. **In a fission reactor, heavy water:**
    (Average Rigor)

    A. Cools off neutrons to control temperature

    B. Moderates fission reactions

    C. Initiates the reaction chain

    D. Dissolves control rods

80. **Given the following values for the masses of a proton, a neutron and an alpha particle, what is the nuclear binding energy of an alpha particle?**
    (Rigorous)

    Proton mass=1.6726×10^{-27} kg
    Neutron mass=1.6749×10^{-27} kg
    Alpha particle mass= 6.6465×10^{-27} kg

    A. 0 J

    B. 7.3417×10^{-27} J

    C. 4 J

    D. 4.3589×10^{-12} J

Answer Key

1.	D	25.	B	49.	B	73.	B
2.	A	26.	C	50.	A	74.	B
3.	B	27.	B	51.	C	75.	B
4.	A	28.	D	52.	A	76.	A
5.	D	29.	C	53.	A	77.	D
6.	C	30.	C	54.	D	78.	D
7.	D	31.	C	55.	C	79.	B
8.	A	32.	D	56.	C	80.	D
9.	B	33.	D	57.	B		
10.	B	34.	C	58.	A		
11.	C	35.	A	59.	D		
12.	D	36.	D	60.	C		
13.	C	37.	C	61.	B		
14.	B	38.	D	62.	B		
15.	A	39.	D	63.	B		
16.	A	40.	B	64.	B		
17.	B	41.	A	65.	A		
18.	C	42.	B	66.	B		
19.	D	43.	D	67.	D		
20.	D	44.	A	68.	C		
21.	B	45.	C	69.	D		
22.	A	46.	C	70.	B		
23.	C	47.	D	71.	C		
24.	C	48.	B	72.	C		

Rigor Analysis Table

Easy	21%	1,8,10,15,20,24,33,35,37,50,52,54,57,59,67,71,75
Average Rigor	39%	2,4,6,7,9,13,14,18,22,27,28,29,30,31,34,38,40,42, 45,47,51,55,56,58,63,64,66,74,76,78,79
Rigorous	40%	3,5,11,12,16,17,19,21,23,25,26,32,36,39,41,43,44, 46,48,49,53,60,61,62,65,68,69,70,72,73,77,80

Rationales with Sample Questions

1. **Which statement best describes a valid approach to testing a scientific hypothesis?**
   (Easy)

   A. Use computer simulations to verify the hypothesis

   B. Perform a mathematical analysis of the hypothesis

   C. Design experiments to test the hypothesis

   D. All of the above

Answer: D

Each of the answers A, B and C can have a crucial part in testing a scientific hypothesis. Although experiments may hold more weight than mathematical or computer-based analysis, these latter two methods of analysis can be critical, especially when experimental design is highly time consuming or financially costly.

2. **Which description best describes the role of a scientific model of a physical phenomenon?**
   (Average Rigor)

   A. An explanation that provides a reasonably accurate approximation of the phenomenon

   B. A theoretical explanation that describes exactly what is taking place

   C. A purely mathematical formulation of the phenomenon

   D. A predictive tool that has no interest in what is actually occurring

Answer: A

A scientific model seeks to provide the most fundamental and accurate description possible for physical phenomena, but, given the fact that natural science takes an *a posteriori* approach, models are always tentative and must be treated with some amount of skepticism. As a result, A is a better answer than B. Answers C and D overly emphasize one or another aspect of a model, rather than a synthesis of a number of aspects (such as a mathematical and predictive aspect).

3. **Which situation calls might best be described as involving an ethical dilemma for a scientist?**
   (Rigorous)

   A. Submission to a peer-review journal of a paper that refutes an established theory

   B. Synthesis of a new radioactive isotope of an element

   C. Use of a computer for modeling a newly-constructed nuclear reactor

   D. Use of a pen-and-paper approach to a difficult problem

Answer: B

Although answer A may be controversial, it does not involve an inherently ethical dilemma, since there is nothing unethical about presenting new information if it is true or valid. Answer C, likewise, has no necessary ethical dimension, as is the case with D. Synthesis of radioactive material, however, involves an ethical dimension with regard to the potential impact of the new isotope on the health of others and on the environment. The potential usefulness of such an isotope in weapons development is another ethical consideration.

4. **Which of the following is not a key purpose for the use of open communication about and peer-review of the results of scientific investigations?**
   (Average Rigor)

   A. Testing, by other scientists, of the results of an investigation for the purpose of refuting any evidence contrary to an established theory

   B. Testing, by other scientists, of the results of an investigation for the purpose of finding or eliminating any errors in reasoning or measurement

   C. Maintaining an open, public process to better promote honesty and integrity in science

   D. Provide a forum to help promote progress through mutual sharing and review of the results of investigations

Answer: A

Answers B, C and D all are important rationales for the use of open communication and peer-review in science. Answer A, however, would suggest that the purpose of these processes is to simply maintain the status quo; the history of science, however, suggests that this cannot and should not be the case.

5. **Which of the following aspects of the use of computers for collecting experimental data is not a concern for the scientist?**
   (Rigorous)

   A. The relative speeds of the processor, peripheral, memory storage unit and any other components included in data acquisition equipment

   B. The financial cost of the equipment, utilities and maintenance

   C. Numerical error due to a lack of infinite precision in digital equipment

   D. The order of complexity of data analysis algorithms

Answer: D

Although answer D might be a concern for later, when actual analysis of the data is undertaken, the collection of data typically does not suffer from this problem. The use of computers does, however, pose problems when, for example, a peripheral collects data at a rate faster than the computer can process it (A), or if the cost of running the equipment or of purchasing the equipment is prohibitive (B). Numerical error is always a concern with any digital data acquisition system, since the data that is collected is never exact.

TEACHER CERTIFICATION STUDY GUIDE

6. **If a particular experimental observation contradicts a theory, what is the most appropriate approach that a physicist should take?**
   (Average Rigor)

   A. Immediately reject the theory and begin developing a new theory that better fits the observed results

   B. Report the experimental result in the literature without further ado

   C. Repeat the observations and check the experimental apparatus for any potential faulty components or human error, and then compare the results once more with the theory

   D. Immediately reject the observation as in error due to its conflict with theory

Answer: C

When experimental results contradict a reigning physical theory, as they do from time to time, it is almost never appropriate to immediately reject the theory (A) *or* the observational results (D). Also, since this is the case, reporting the result in the literature, without further analysis to provide an adequate explanation of the discrepancy, is unwise and unwarranted. Further testing is appropriate to determine whether the experiment is repeatable and whether any equipment or human errors have occurred. Only after further testing may the physicist begin to analyze the implications of the observational result.

7. **Which of the following is *not* an SI unit?**
   (Average Rigor)

   A. Joule

   B. Coulomb

   C. Newton

   D. Erg

Answer: D

The first three responses are the SI (*Le Système International d'Unités*) units for energy, charge and force, respectively. The fourth answer, the erg, is the CGS (centimeter-gram-second) unit of energy.

8. **Which of the following best describes the relationship of precision and accuracy in scientific measurements?**
   (Easy)

   A. Accuracy is how well a particular measurement agrees with the value of the actual parameter being measured; precision is how well a particular measurement agrees with the average of other measurements taken for the same value

   B. Precision is how well a particular measurement agrees with the value of the actual parameter being measured; accuracy is how well a particular measurement agrees with the average of other measurements taken for the same value

   C. Accuracy is the same as precision

   D. Accuracy is a measure of numerical error; precision is a measure of human error

Answer: A

The accuracy of a measurement is how close the measurement is to the "true" value of the parameter being measured. Precision is how closely a group of measurements is to the mean value of all the measurements. By analogy, accuracy is how close a measurement is to the center of the bulls-eye, and precision is how tight a group is formed by multiple measurements, regardless of accuracy. Thus, measurements may be very precise and not very accurate, or they may be accurate but not overly precise, or they may be both or neither.

9. **Which statement best describes a rationale for the use of statistical analysis in characterizing the numerical results of a scientific experiment or investigation?**
   (Average Rigor)

   A. Experimental results need to be adjusted, through the use of statistics, to conform to theoretical predictions and computer models

   B. Since experiments are prone to a number of errors and uncertainties, statistical analysis provides a method for characterizing experimental measurements by accounting for or quantifying these undesirable effects

   C. Experiments are not able to provide any useful information, and statistical analysis is needed to impose a theoretical framework on the results

   D. Statistical analysis is needed to relate experimental measurements to computer-simulated values

Answer: B

One of the main reasons for the use of statistical analysis is that various types of noise, errors and uncertainties can easily enter into experimental results. Among other things, statistics can help alleviate these difficulties by quantifying an average measurement value and a variance or standard deviation of the set of measurements. This helps determine the accuracy and precision of a set of experimental results. Answers A, C and D do not accurately describe ideal scientific experiments or the use of statistics.

10. **Which statement best characterizes the relationship of mathematics and experimentation in physics?**
    (Easy)

    A. Experimentation has no bearing on the mathematical models that are developed for physical phenomena

    B. Mathematics is a tool that assists in the development of models for various physical phenomena as they are studied experimentally, with observations of the phenomena being a test of the validity of the mathematical model

    C. Mathematics is used to test the validity of experimental apparatus for physical measurements

    D. Mathematics is an abstract field with no relationship to concrete experimentation

Answer: B

Mathematics is used extensively in the study of physics for creating models of various phenomena. Since mathematics is abstract and not necessarily tied to physical reality, it must be tempered by experimental results. Although a particular theory may be mathematically elegant, it may have no explanatory power due to its inability to account for certain aspects of physical reality, or due to its inclusion of gratuitous aspects that seem to have no physical analog. Thus, experimentation is foundational, with mathematics being a tool for organizing and providing a greater context for observational results.

11. **Which of the following mathematical tools would not typically be used for the analysis of an electromagnetic phenomenon?**
    (Rigorous)

    A. Trigonometry

    B. Vector calculus

    C. Group theory

    D. Numerical methods

Answer: C

Trigonometry and vector calculus are both key tools for solving problems in electromagnetics. These are, primarily, analytical methods, although they play a part in numerical analysis as well. Numerical methods are helpful for many problems that are otherwise intractable analytically. Group theory, although it may have some applications in certain highly specific areas, is generally not used in the study of electromagnetics.

12. For a problem that involves parameters that vary in rate with direction and location, which of the following mathematical tools would most likely be of greatest value?
    (Rigorous)

    A. Trigonometry

    B. Numerical analysis

    C. Group theory

    D. Vector calculus

Answer: D

Each of the above answers might have some value for individual problems, but, generally speaking, those problems that deal with quantities that have direction and magnitude (vectors), and that deal with rates, would most likely be amenable to analysis using vector calculus (D).

13. Which of the following devices would be best suited for an experiment designed to measure alpha particle emissions from a sample?
    (Average Rigor)

    A. Photomultiplier tube

    B. Thermocouple

    C. Geiger-Müller tube

    D. Transistor

Answer: C

The Geiger-Müller tube is the main component of the so-called Geiger counter, which is designed specifically for detecting ionizing radiation emissions, including alpha particles. The photomultiplier tube is better suited to measurement of electromagnetic radiation closer to the visible range (A), and the thermocouple is better suited to measurement of temperature (B). Transistors may be involved in instrumentation, but they are not sensors.

14. **Which of the following experiments presents the most likely cause for concern about laboratory safety?**
    (Average Rigor)

    A. Computer simulation of a nuclear reactor

    B. Vibration measurement with a laser

    C. Measurement of fluorescent light intensity with a battery-powered photodiode circuit

    D. Ambient indoor ionizing radiation measurement with a Geiger counter.

Answer: B

Assuming no profoundly foolish acts, the use of a computer for simulation (A), measurement with a battery-powered photodiode circuit (C) and ambient radiation measurement (D) pose no particular hazards. The use of a laser (B) must be approached with care, however, as unintentional reflections or a lack of sufficient protection can cause permanent eye damage.

15. **A brick and hammer fall from a ledge at the same time. They would be expected to:**
    (Easy)

    A. Reach the ground at the same time

    B. Accelerate at different rates due to difference in weight

    C. Accelerate at different rates due to difference in potential energy

    D. Accelerate at different rates due to difference in kinetic energy

Answer: A

This is a classic question about falling in a gravitational field. All objects are acted upon equally by gravity, so they should reach the ground at the same time. (In real life, air resistance can make a difference, but not at small heights for similarly shaped objects.) In any case, weight, potential energy, and kinetic energy do not affect gravitational acceleration. Thus, the only possible answer is (A).

16. A baseball is thrown with an initial velocity of 30 m/s at an angle of 45°. Neglecting air resistance, how far away will the ball land?
(Rigorous)

    A. 92 m

    B. 78 m

    C. 65 m

    D. 46 m

Answer: A

To answer this question, recall the equations for projectile motion:
$y = \frac{1}{2} a t^2 + v_{0y} t + y_0$
$x = v_{0x} t + x_0$
where x and y are horizontal and vertical position, respectively; t is time; a is acceleration due to gravity; v_{0x} and v_{0y} are initial horizontal and vertical velocity, respectively; x_0 and y_0 are initial horizontal and vertical position, respectively.
For our case:
x_0 and y_0 can be set to zero
both v_{0x} and v_{0y} are (using trigonometry) = $(\sqrt{2}/2)$ 30 m/s
$a = -9.81$ m/s^2

We then use the vertical motion equation to find the time aloft (setting y equal to zero to find the solution for t):
$0 = \frac{1}{2} (-9.81 \text{ m/s}^2) t^2 + (\sqrt{2}/2)$ 30 m/s t
Then solving, we find:
t = 0 s (initial set-up) or t = 4.324 s (time to go up and down)

Using t = 4.324 s in the horizontal motion equation, we find:
$x = ((\sqrt{2}/2)$ 30 m/s$)$ (4.324 s)
x = 91.71 m

This is consistent only with answer (A).

17. A skateboarder accelerates down a ramp, with constant acceleration of two meters per second squared, from rest. The distance in meters, covered after four seconds, is:
   (Rigorous)

   A. 10

   B. 16

   C. 23

   D. 37

Answer: B

To answer this question, recall the equation relating constant acceleration to distance and time:
$x = \frac{1}{2} a t^2 + v_0 t + x_0$ where x is position; a is acceleration; t is time; v_0 and x_0 are initial velocity and position (both zero in this case)

thus, to solve for x:
$x = \frac{1}{2} (2 \text{ m/s}^2)(4^2 \text{s}^2) + 0 + 0$
$x = 16$ m

This is consistent only with answer (B).

18. When acceleration is plotted versus time, the area under the graph represents:
   (Average Rigor)

   A. Time

   B. Distance

   C. Velocity

   D. Acceleration

Answer: C
The area under a graph will have units equal to the product of the units of the two axes. (To visualize this, picture a graphed rectangle with its area equal to length times width.)
Therefore, multiply units of acceleration by units of time:
(length/time^2)(time)
This equals length/time, i.e. units of velocity.

19. An inclined plane is tilted by gradually increasing the angle of elevation θ, until the block will slide down at a constant velocity. The coefficient of friction, μ_k, is given by:
   (Rigorous)

   A. cos θ

   B. sin θ

   C. cosecant θ

   D. tangent θ

Answer: D

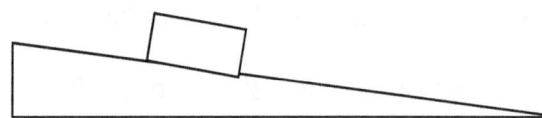

When the block moves, its force upstream (due to friction) must equal its force downstream (due to gravity).

The friction force is given by
$F_f = \mu_k N$
where μ_k is the friction coefficient and N is the normal force.

Using similar triangles, the gravity force is given by
$F_g = mg \sin θ$
and the normal force is given by
$N = mg \cos θ$

When the block moves at constant velocity, it must have zero net force, so set equal the force of gravity and the force due to friction:
$F_f = F_g$
$\mu_k mg \cos θ = mg \sin θ$
$\mu_k = \tan θ$

Answer (D) is the only appropriate choice in this case.

20. An object traveling through air loses part of its energy of motion due to friction. Which statement best describes what has happened to this energy?
   (Easy)

   A. The energy is destroyed

   B. The energy is converted to static charge

   C. The energy is radiated as electromagnetic waves

   D. The energy is lost to heating of the air

Answer: D

Since energy must be conserved, the energy of motion of the object is converted, in part, to energy of motion of the molecules in the air (and, to some extent, in the object). This additional motion is equivalent to an increase in heat. Thus, friction is a loss of energy of motion through heating.

21. The weight of an object on the earth's surface is designated x. When it is two earth's radii from the surface of the earth, its weight will be:
   (Rigorous)

   A. $x/4$

   B. $x/9$

   C. $4x$

   D. $16x$

Answer: B

To solve this problem, apply the universal Law of Gravitation to the object and Earth:

$F_{gravity} = (GM_1M_2)/R^2$

Because the force of gravity varies with the square of the radius between the objects, the force (or weight) on the object will be decreased by the square of the multiplication factor on the radius. Note that the object on Earth's surface is *already* at one radius from Earth's center. Thus, when it is two radii from Earth's surface, it is three radii from Earth's center. R^2 is then nine, so the weight is $x/9$. Only answer (B) matches these calculations.

PHYSICS

22. Which of the following units is not used to measure torque?
(Average Rigor)

    A. slug ft

    B. lb ft

    C. N m

    D. dyne cm

Answer: A

To answer this question, recall that torque is always calculated by multiplying units of force by units of distance. Therefore, answer (A), which is the product of units of mass and units of distance, must be the choice of incorrect units. Indeed, the other three answers all could measure torque, since they are of the correct form. It is a good idea to review "English Units" before the teacher test, because they are occasionally used in problems.

23. A uniform pole weighing 100 grams, that is one meter in length, is supported by a pivot at 40 centimeters from the left end. In order to maintain static position, a 200 gram mass must be placed _____ centimeters from the left end.
(Rigorous)

    A. 10

    B. 45

    C. 35

    D. 50

Answer: C

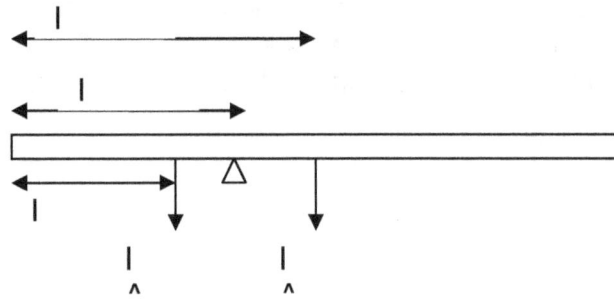

Since the pole is uniform, we can assume that its weight 0.1g acts at the center, i.e. 50 cm from the left end. In order to keep the pole balanced on the pivot, the 200 gram mass must be placed such that the torque on the pole due to the mass is equal and opposite to the torque due to the pole's weight. Thus, if the 200 gram mass is placed d cm from the left end of the pole,

$(40 - d) \times 0.2g = 10 \times 0.1g$; $40 - d = 5$; $d = 35$ cm

24. The magnitude of a force is:
(Easy)

    A. Directly proportional to mass and inversely to acceleration

    B. Inversely proportional to mass and directly to acceleration

    C. Directly proportional to both mass and acceleration

    D. Inversely proportional to both mass and acceleration

Answer: C

To solve this problem, recall Newton's 2nd Law, i.e. net force is equal to mass times acceleration. Therefore, the only possible answer is (C).

25. A projectile with a mass of 1.0 kg has a muzzle velocity of 1500.0 m/s when it is fired from a cannon with a mass of 500.0 kg. If the cannon slides on a frictionless track, it will recoil with a velocity of ____ m/s.
(Rigorous)

    A. 2.4

    B. 3.0

    C. 3.5

    D. 1500

Answer: B

To solve this problem, apply Conservation of Momentum to the cannon-projectile system. The system is initially at rest, with total momentum of 0 kg m/s. Since the cannon slides on a frictionless track, we can assume that the net momentum stays the same for the system. Therefore, the momentum forward (of the projectile) must equal the momentum backward (of the cannon). Thus:

    $p_{projectile} = p_{cannon}$
    $m_{projectile} \, v_{projectile} = m_{cannon} \, v_{cannon}$
    (1.0 kg)(1500.0 m/s) = (500.0 kg)(x)
    x = 3.0 m/s

Only answer (B) matches these calculations.

26. A car (mass m_1) is driving at velocity v, when it smashes into an unmoving car (mass m_2), locking bumpers. Both cars move together at the same velocity. The common velocity will be given by: *(Rigorous)*

   A. $m_1 v/m_2$

   B. $m_2 v/m_1$

   C. $m_1 v/(m_1 + m_2)$

   D. $(m_1 + m_2)v/m_1$

Answer: C

In this problem, there is an inelastic collision, so the best method is to assume that momentum is conserved. (Recall that momentum is equal to the product of mass and velocity.)
Therefore, apply Conservation of Momentum to the two-car system:
Momentum at Start = Momentum at End
(Mom. of Car 1) + (Mom. of Car 2) = (Mom. of 2 Cars Coupled)
$m_1 v + 0 = (m_1 + m_2)x$
$x = m_1 v/(m_1 + m_2)$
Only answer (C) matches these calculations.

Watch out for the other answers, because errors in algebra could lead to a match with incorrect answer (D), and assumption of an elastic collision could lead to a match with incorrect answer (A).

27. A satellite is in a circular orbit above the earth. Which statement is false?
    (Average Rigor)

    A. An external force causes the satellite to maintain orbit.

    B. The satellite's inertia causes it to maintain orbit.

    C. The satellite is accelerating toward the earth.

    D. The satellite's velocity and acceleration are not in the same direction.

Answer: B

To answer this question, recall that in circular motion, an object's inertia tends to keep it moving straight (tangent to the orbit), so a centripetal force (leading to centripetal acceleration) must be applied. In this case, the centripetal force is gravity due to the earth, which keeps the object in motion. Thus, (A), (C), and (D) are true, and (B) is the only false statement.

28. A 100 g mass revolving around a fixed point, on the end of a 0.5 meter string, circles once every 0.25 seconds. What is the magnitude of the centripetal acceleration?
    (Average Rigor)

    A. 1.23 m/s^2

    B. 31.6 m/s^2

    C. 100 m/s^2

    D. 316 m/s^2

Answer: D

The centripetal acceleration is equal to the product of the radius and the square of the angular frequency ω. In this case, ω is equal to 25.1 Hz. Squaring this value and multiplying by 0.5 m yields the result in answer D.

29. Which statement best describes the relationship of simple harmonic motion to a simple pendulum of length L, mass m and displacement of arc length s?
(Average Rigor)

    A. A simple pendulum cannot be modeled using simple harmonic motion

    B. A simple pendulum may be modeled using the same expression as Hooke's law for displacement s, but with a spring constant equal to the tension on the string

    C. A simple pendulum may be modeled using the same expression as Hooke's law for displacement s, but with a spring constant equal to m g/L

    D. A simple pendulum typically does not undergo simple harmonic motion

Answer: C

The force on a simple pendulum may be expressed approximately (when displacement s is small) according to the following equation:

$$F \approx -\frac{mg}{L}s$$

This expression has the same form as Hooke's law (F = -kx). Thus, answer C is the most correct response. Another approach to the question is to eliminate answers A and D as obviously incorrect, and then to eliminate answer B as not having appropriate units for the spring constant.

30. A mass of 2 kg connected to a spring undergoes simple harmonic motion at a frequency of 3 Hz. What is the spring constant?
(Average Rigor)

    A. 6 kg/s^2

    B. 18 kg/s^2

    C. 710 kg/s^2

    D. 1000 kg/s^2

Answer: C

The spring constant, k, is equal to mω^2. In this case, ω is equal to 2π times the frequency of 3 Hz. The spring constant may be derived quickly by recognizing that the position of the mass varies sinusoidally with time at an angular frequency ω. Noting that the acceleration is the second derivative of the position with respect to time, the expression for k in Hooke's law (F = -kx) can be easily derived.

31. The kinetic energy of an object is _____ proportional to its _____.
(Average Rigor)

    A. Inversely...inertia

    B. Inversely...velocity

    C. Directly...mass

    D. Directly...time

Answer: C

To answer this question, recall that kinetic energy is equal to one-half of the product of an object's mass and the square of its velocity:
KE = ½ m v^2

Therefore, kinetic energy is directly proportional to mass, and the answer is (C). Note that although kinetic energy is associated with both velocity and momentum (a measure of inertia), it is not *inversely* proportional to either one.

TEACHER CERTIFICATION STUDY GUIDE

32. A force is given by the vector 5 N x + 3 N y (where x and y are the unit vectors for the x- and y- axes, respectively). This force is applied to move a 10 kg object 5 m, in the x direction. How much work was done?
    (Rigorous)

    A. 250 J

    B. 400 J

    C. 40 J

    D. 25 J

Answer: D

To find out how much work was done, note that work counts only the force in the direction of motion. Therefore, the only part of the vector that we use is the 5 N in the x-direction. Note, too, that the mass of the object is not relevant in this problem. We use the work equation:
Work = (Force in direction of motion) (Distance moved)
Work = (5 N) (5 m)
Work = 25 J
This is consistent only with answer (D).

33. An office building entry ramp uses the principle of which simple machine?
    (Easy)

    A. Lever

    B. Pulley

    C. Wedge

    D. Inclined Plane

Answer: D

To answer this question, recall the definitions of the various simple machines. A ramp, which trades a longer traversed distance for a shallower slope, is an example of an Inclined Plane, consistent with answer (D). Levers and Pulleys act to change size and/or direction of an input force, which is not relevant here. Wedges apply the same force over a smaller area, increasing pressure—again, not relevant in this case.

PHYSICS

34. If the internal energy of a system remains constant, how much work is done by the system if 1 kJ of heat energy is added?
(Average Rigor)

    A. 0 kJ

    B. -1 kJ

    C. 1 kJ

    D. 3.14 kJ

Answer: C

According to the first law of thermodynamics, if the internal energy of a system remains constant, then any heat energy added to the system must be balanced by the system performing work on its surroundings. In the case of an ideal gas, the gas would necessarily expand when heated, assuming a constant internal energy was somehow maintained. Applying conservation of energy, answer C is found to be correct.

35. A calorie is the amount of heat energy that will:
(Easy)

    A. Raise the temperature of one gram of water from 14.5° C to 15.5° C.

    B. Lower the temperature of one gram of water from 16.5° C to 15.5° C

    C. Raise the temperature of one gram of water from 32° F to 33° F

    D. Cause water to boil at two atmospheres of pressure.

Answer: A

The definition of a calorie is, "the amount of energy to raise one gram of water by one degree Celsius," and so answer (A) is correct. Do not get confused by the fact that 14.5° C seems like a random number. Also, note that answer (C) tries to confuse you with degrees Fahrenheit, which are irrelevant to this problem.

36. Use the information on heats below to solve this problem. An ice block at 0° Celsius is dropped into 100 g of liquid water at 18° Celsius. When thermal equilibrium is achieved, only liquid water at 0° Celsius is left. What was the mass, in grams, of the original block of ice?

   Given: Heat of fusion of ice = 80 cal/g
   Heat of vaporization of ice = 540 cal/g
   Specific Heat of ice = 0.50 cal/g°C
   Specific Heat of water = 1 cal/g°C

   (Rigorous)

   A. 2.0

   B. 5.0

   C. 10.0

   D. 22.5

Answer: D

To solve this problem, apply Conservation of Energy to the ice-water system. Any gain of heat to the melting ice must be balanced by loss of heat in the liquid water. Use the two equations relating temperature, mass, and energy:
Q = m C ΔT (for heat loss/gain from change in temperature)
Q = m L (for heat loss/gain from phase change)
where Q is heat change; m is mass; C is specific heat; ΔT is change in temperature; L is heat of phase change (in this case, melting, also known as "fusion").

Then
$Q_{ice\ to\ water}$ = $Q_{water\ to\ ice}$
(Note that the ice only melts; it stays at 0° Celsius—otherwise, we would have to include a term for warming the ice as well. Also the information on the heat of vaporization for water is irrelevant to this problem.)
m L = m C ΔT
x (80 cal/g) = 100g 1cal/g°C 18°C
x (80 cal/g) = 1800 cal
x = 22.5 g

   Only answer (D) matches this result.

TEACHER CERTIFICATION STUDY GUIDE

37. **Heat transfer by electromagnetic waves is termed:**
    (Easy)

    A. Conduction

    B. Convection

    C. Radiation

    D. Phase Change

Answer: C

To answer this question, recall the different ways that heat is transferred. Conduction is the transfer of heat through direct physical contact and molecules moving and hitting each other. Convection is the transfer of heat via density differences and flow of fluids. Radiation is the transfer of heat via electromagnetic waves (and can occur in a vacuum). Phase Change causes transfer of heat (though not of temperature) in order for the molecules to take their new phase. This is consistent, therefore, only with answer (C).

38. **A cooking thermometer in an oven works because the metals it is composed of have different:**
    (Average Rigor)

    A. Melting points

    B. Heat convection

    C. Magnetic fields

    D. Coefficients of expansion

Answer: D

A thermometer of the type that can withstand oven temperatures works by having more than one metal strip. These strips expand at different rates with temperature increases, causing the dial to register the new temperature. This is consistent only with answer (D). If you did not know how an oven thermometer works, you could still omit the incorrect answers: It is unlikely that the metals in a thermometer would melt in the oven to display the temperature; the magnetic fields would not be useful information in this context; heat convection applies in fluids, not solids.

PHYSICS

39. **Which of the following is not an assumption upon which the kinetic-molecular theory of gases is based?**
    (Rigorous)

    A. Quantum mechanical effects may be neglected

    B. The particles of a gas may be treated statistically

    C. The particles of the gas are treated as very small masses

    D. Collisions between gas particles and container walls are inelastic

Answer: D

Since the kinetic-molecular theory is classical in nature, quantum mechanical effects are indeed ignored, and answer A is incorrect. The theory also treats gases as a statistical collection of point-like particles with finite masses. As a result, answers B and C may also be eliminated. Thus, answer D is correct: collisions between gas particles and container walls are treated as elastic in the kinetic-molecular theory.

40. **What is temperature?**
    (Average Rigor)

    A. Temperature is a measure of the conductivity of the atoms or molecules in a material

    B. Temperature is a measure of the kinetic energy of the atoms or molecules in a material

    C. Temperature is a measure of the relativistic mass of the atoms or molecules in a material

    D. Temperature is a measure of the angular momentum of electrons in a material

Answer: B

Temperature is, in fact, a measure of the kinetic energy of the constituent components of a material. Thus, as a material is heated, the atoms or molecules that compose it acquire greater energy of motion. This increased motion results in the breaking of chemical bonds and in an increase in disorder, thus leading to melting or vaporizing of the material at sufficiently high temperatures.

41. **Solids expand when heated because:**
(Rigorous)

    A. Molecular motion causes expansion

    B. PV = nRT

    C. Magnetic forces stretch the chemical bonds

    D. All material is effectively fluid

Answer: A

When any material is heated, the heat energy becomes energy of motion for the material's molecules. This increased motion causes the material to expand (or sometimes to change phase). Therefore, the answer is (A). Answer (B) is the ideal gas law, which gives a relationship between temperature, pressure, and volume for gases. Answer (C) is a red herring (misleading answer that is untrue). Answer (D) may or may not be true, but it is not the best answer to this question.

42. What should be the behavior of an electroscope, which has been grounded in the presence of a positively charged object (1), after the ground connection is removed and then the charged object is removed from the vicinity (2)?
(Average Rigor)

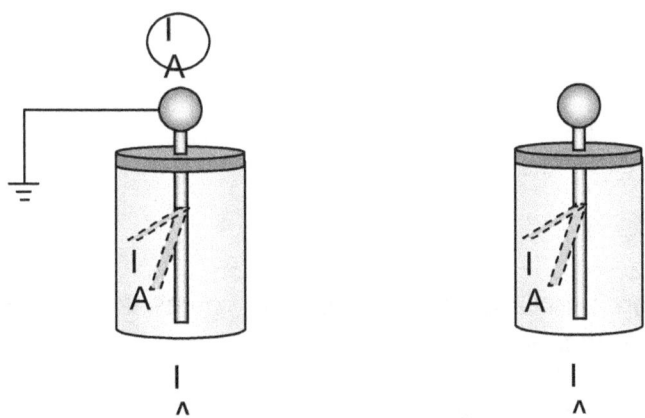

A. The metal leaf will start deflected (1) and then relax to an undeflected position (2)

B. The metal leaf will start in an undeflected position (1) and then be deflected (2)

C. The metal leaf will remain undeflected in both cases

D. The metal leaf will be deflected in both cases

Answer: B

When grounded, the electroscope will show no deflection. Nevertheless, if the ground is then removed and the charged object taken from the vicinity (in that order), the excess charge that existed near the sphere of the electroscope will distribute itself throughout the instrument, resulting in an overall net excess charge that will deflect the metal leaf.

43. The electric force in Newtons, on two small objects (each charged to −10 microCoulombs and separated by 2 meters) is:
(Rigorous)

    A. 1.0

    B. 9.81

    C. 31.0

    D. 0.225

Answer: D

To answer this question, use Coulomb's Law, which gives the electric force between two charged particles:
$F = k Q_1 Q_2 / r^2$
Then our unknown is F, and our knowns are:
$k = 9.0 \times 10^9 \text{ Nm}^2/\text{C}^2$
$Q_1 = Q_2 = -10 \times 10^{-6} \text{ C}$
$r = 2 \text{ m}$

Therefore
$F = (9.0 \times 10^9)(-10 \times 10^{-6})(-10 \times 10^{-6})/(2^2)$ N
$F = 0.225$ N

This is compatible only with answer (D).

44. A 10 ohm resistor and a 50 ohm resistor are connected in parallel. If the current in the 10 ohm resistor is 5 amperes, the current (in amperes) running through the 50 ohm resistor is:
(Rigorous)

    A. 1

    B. 50

    C. 25

    D. 60

Answer: A

To answer this question, use Ohm's Law, which relates voltage to current and resistance:
$V = IR$
where V is voltage; I is current; R is resistance.

We also use the fact that in a parallel circuit, the voltage is the same across the branches.

Because we are given that in one branch, the current is 5 amperes and the resistance is 10 ohms, we deduce that the voltage in this circuit is their product, 50 volts (from $V = IR$).

We then use $V = IR$ again, this time to find I in the second branch. Because V is 50 volts, and R is 50 ohm, we calculate that I has to be 1 ampere.

This is consistent only with answer (A).

45. How much power is dissipated through the following resistive circuit?
(Average Rigor)

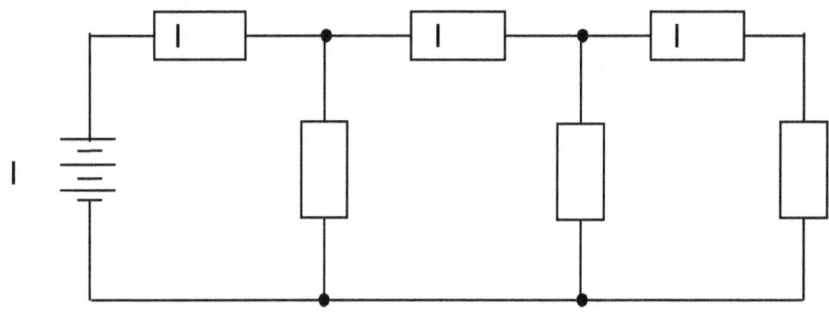

A. 0 W

B. 0.22 W

C. 0.31 W

D. 0.49 W

Answer: C

Use the rules of series and parallel resistors to quickly form an equivalent circuit with a single voltage source and a single resistor. In this case, the equivalent resistance is 3.25 Ω. The power dissipated by the circuit is the square of the voltage divided by the resistance. The final answer is C.

46. **The greatest number of 100 watt lamps that can be connected in parallel with a 120 volt system without blowing a 5 amp fuse is:**
(Rigorous)

   A. 24

   B. 12

   C. 6

   D. 1

Answer: C

To solve fuse problems, you must add together all the drawn current in the parallel branches, and make sure that it is less than the fuse's amp measure. Because we know that electrical power is equal to the product of current and voltage, we can deduce that:
I = P/V (I = current (amperes); P = power (watts); V = voltage (volts))

Therefore, for each lamp, the current is 100/120 amperes, or 5/6 ampere. The highest possible number of lamps is thus six, because six lamps at 5/6 ampere each adds to 5 amperes; more will blow the fuse.

This is consistent only with answer (C).

47. **Which of the following statements may be taken as a legitimate inference based upon the Maxwell equation that states $\nabla \cdot \mathbf{B} = 0$?**
(Average Rigor)

   A. The electric and magnetic fields are decoupled

   B. The electric and magnetic fields are mediated by the W boson

   C. There are no photons

   D. There are no magnetic monopoles

Answer: D

Since the divergence of the magnetic flux density is always zero, there cannot be any magnetic monopoles (charges), given this Maxwell equation. If Gauss's law is applied to magnetic flux in the same manner as it is to electric flux, then the total magnetic "charge" contained within any closed surface must always be zero. This is another way of viewing the problem. Thus, answer D is correct. This answer may also be chosen by elimination of the other statements, which are untenable.

48. What effect might an applied external magnetic field have on the magnetic domains of a ferromagnetic material?
(Rigorous)

    A. The domains that are not aligned with the external field increase in size, but those that are aligned decrease in size

    B. The domains that are not aligned with the external field decrease in size, but those that are aligned increase in size

    C. The domains align perpendicular to the external field

    D. There is no effect on the magnetic domains

Answer: B

Recall that ferromagnetic domains are portions of a magnetic material that have a local magnetic moment. The material may have an overall lack of a magnetic moment due to random alignment of its domains. In the presence of an applied field, the domains may align with the field to some extent, or the boundaries of the domains may shift to give greater weight to those domains that are aligned with the field, at the expense of those domains that are not aligned with the field. As a result, of the possibilities above, B is the best answer.

49. What is the effect of running current in the same direction along two parallel wires, as shown below?
   (Rigorous)

   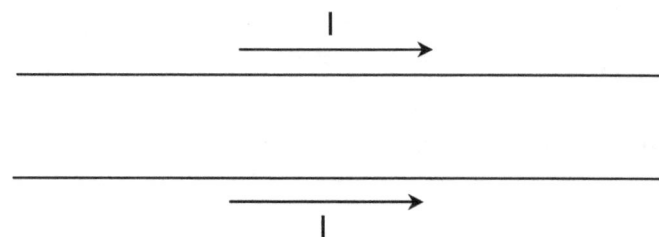

   A. There is no effect

   B. The wires attract one another

   C. The wires repel one another

   D. A torque is applied to both wires

Answer: B

Since the direction of the force on a current element is proportional to the cross product of the direction of the current element and the magnetic field, there is either an attractive or repulsive force between the two wires shown above. Using the right hand rule, it can be found that the magnetic field on the top wire due to the bottom wire is directed out of the plane of the page. Performing the cross product shows that the force on the upper wire is directed toward the lower wire. A similar argument can be used for the lower wire. Thus, the correct answer is B: an attractive force is exerted on the wires.

TEACHER CERTIFICATION STUDY GUIDE

50. **The current induced in a coil is defined by which of the following laws?**
(Easy)

    A. Lenz's Law

    B. Burke's Law

    C. The Law of Spontaneous Combustion

    D. Snell's Law

Answer: A

Lenz's Law states that an induced electromagnetic force always gives rise to a current whose magnetic field opposes the original flux change. There is no relevant "Snell's Law," "Burke's Law," or "Law of Spontaneous Combustion" in electromagnetism. (In fact, only Snell's Law is a real law of these three, and it refers to refracted light.) Therefore, the only appropriate answer is (A).

51. **A light bulb is connected in series with a rotating coil within a magnetic field. The brightness of the light may be increased by any of the following except:**
(Average Rigor)

    A. Rotating the coil more rapidly.

    B. Using more loops in the coil.

    C. Using a different color wire for the coil.

    D. Using a stronger magnetic field.

Answer: C

To answer this question, recall that the rotating coil in a magnetic field generates electric current, by Faraday's Law. Faraday's Law states that the amount of emf generated is proportional to the rate of change of magnetic flux through the loop. This increases if the coil is rotated more rapidly (A), if there are more loops (B), or if the magnetic field is stronger (D). Thus, the only answer to this question is (C).

52. **What is the direction of the magnetic field at the center of the loop of current (I) shown below (i.e., at point A)?**
    (Easy)

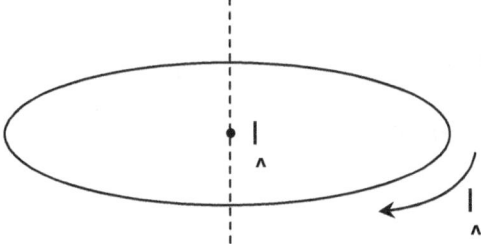

    A. Down, along the axis (dotted line)

    B. Up, along the axis (dotted line)

    C. The magnetic field is oriented in a radial direction

    D. There is no magnetic field at point A

Answer: A

The magnetic field may be found by applying the right-hand rule. The magnetic field curls around the wire in the direction of the curled fingers when the thumb is pointed in the direction of the current. Since there is a degree of symmetry, with point A lying in the center of the loop, the contributions of all the current elements on the loop must yield a field that is either directed up or down at the axis. Use of the right-hand rule indicates that the field is directed down. Thus, answer A is correct.

TEACHER CERTIFICATION STUDY GUIDE

53. **The use of two circuits next to each other, with a change in current in the primary circuit, demonstrates:**
    (Rigorous)

    A. Mutual current induction

    B. Dielectric constancy

    C. Harmonic resonance

    D. Resistance variation

Answer: A

To answer this question, recall that changing current induces a change in magnetic flux, which in turn causes a change in current to oppose that change (Lenz's and Faraday's Laws). Thus, (A) is correct. If you did not remember that, note that harmonic resonance is irrelevant here (eliminating (C)), and there is no change in resistance in the circuits (eliminating (D)).

54. **A semi-conductor allows current to flow:**
    (Easy)

    A. Never

    B. Always

    C. As long as it stays below a maximum temperature

    D. When a minimum voltage is applied

Answer: D

To answer this question, recall that semiconductors do not conduct as well as conductors (eliminating answer (B)), but they conduct better than insulators (eliminating answer (A)). Semiconductors can conduct better when the temperature is higher (eliminating answer (C)), and their electrons move most readily under a potential difference. Thus the answer can only be (D).

PHYSICS

TEACHER CERTIFICATION STUDY GUIDE

55. All of the following use semi-conductor technology, except a(n):
(Average Rigor)

    A. Transistor

    B. Diode

    C. Capacitor

    D. Operational Amplifier

Answer: C

Semi-conductor technology is used in transistors and operational amplifiers, and diodes are the basic unit of semi-conductors. Therefore the only possible answer is (C), and indeed a capacitor does not require semi-conductor technology.

56. A wave generator is used to create a succession of waves. The rate of wave generation is one every 0.33 seconds. The period of these waves is:
(Average Rigor)

    A. 2.0 seconds

    B. 1.0 seconds

    C. 0.33 seconds

    D. 3.0 seconds

Answer: C

The definition of a period is the length of time between wave crests. Therefore, when waves are generated one per 0.33 seconds, that same time (0.33 seconds) is the period. This is consistent only with answer (C). Do not be trapped into calculating the number of waves per second, which might lead you to choose answer (D).

TEACHER CERTIFICATION STUDY GUIDE

57. An electromagnetic wave propagates through a vacuum. Independent of its wavelength, it will move with constant:
(Easy)

    A. Acceleration

    B. Velocity

    C. Induction

    D. Sound

Answer: B

Electromagnetic waves are considered always to travel at the speed of light, so answer (B) is correct. Answers (C) and (D) can be eliminated in any case, because induction is not relevant here, and sound does not travel in a vacuum.

58. A wave has speed 60 m/s and wavelength 30,000 m. What is the frequency of the wave?
(Average Rigor)

    A. 2.0×10^{-3} Hz

    B. 60 Hz

    C. 5.0×10^{2} Hz

    D. 1.8×10^{6} Hz

Answer: A

To answer this question, recall that wave speed is equal to the product of wavelength and frequency. Thus:
60 m/s = (30,000 m) (frequency)
frequency = 2.0×10^{-3} Hz

This is consistent only with answer (A).

PHYSICS

59. Rainbows are created by:
(Easy)

    A. Reflection, dispersion, and recombination

    B. Reflection, resistance, and expansion

    C. Reflection, compression, and specific heat

    D. Reflection, refraction, and dispersion

Answer: D

To answer this question, recall that rainbows are formed by light that goes through water droplets and is dispersed into its colors. This is consistent with both answers (A) and (D). Then note that refraction is important in bending the differently colored light waves, while recombination is not a relevant concept here. Therefore, the answer is (D).

60. Which of the following is *not* a legitimate explanation for refraction of light rays at boundaries between different media?
(Rigorous)

    A. Light seeks the path of least time between two different points

    B. Due to phase matching and other boundary conditions, plane waves travel in different directions on either side of the boundary, depending on the material parameters

    C. The electric and magnetic fields become decoupled at the boundary

    D. Light rays obey Snell's law

Answer: C

Even if the exact implications of each explanation are not known or understood, answer C can be chosen due to its plain incorrectness. The other responses involve more or less fundamental explanations for the refraction of light rays (which are equivalent to plane waves) at media boundaries.

61. A stationary sound source produces a wave of frequency *F*. An observer at position A is moving toward the horn, while an observer at position B is moving away from the horn. Which of the following is true?
(Rigorous)

    A. $F_A < F < F_B$

    B. $F_B < F < F_A$

    C. $F < F_A < F_B$

    D. $F_B < F_A < F$

Answer: B

To answer this question, recall the Doppler Effect. As a moving observer approaches a sound source, s/he intercepts wave fronts sooner than if s/he were standing still. Therefore, the wave fronts seem to be coming more frequently. Similarly, as an observer moves away from a sound source, the wave fronts take longer to reach him/her. Therefore, the wave fronts seem to be coming less frequently. Because of this effect, the frequency at B will seem lower than the original frequency, and the frequency at A will seem higher than the original frequency. The only answer consistent with this is (B). Note also, that even if you weren't sure of which frequency should be greater/smaller, you could still reason that A and B should have opposite effects, and be able to eliminate answer choices (C) and (D).

62. A monochromatic ray of light passes from air to a thick slab of glass (n = 1.41) at an angle of 45° from the normal. At what angle does it leave the air/glass interface?
    (Rigorous)

    A. 45°

    B. 30°

    C. 15°

    D. 55°

Answer: B

To solve this problem use Snell's Law:
$n_1 \sin\theta_1 = n_2 \sin\theta_2$ (where n_1 and n_2 are the indexes of refraction and θ_1 and θ_2 are the angles of incidence and refraction).

Then, since the index of refraction for air is 1.0, we deduce:
$1 \sin 45° = 1.41 \sin x$
$x = \sin^{-1}((1/1.41) \sin 45°)$
$x = 30°$

This is consistent only with answer (B). Also, note that you could eliminate answers (A) and (D) in any case, because the refracted light will have to bend at a smaller angle when entering glass.

63. If one sound is ten decibels louder than another, the ratio of the intensity of the first to the second is:
(Average Rigor)

    A. 20:1

    B. 10:1

    C. 1:1

    D. 1:10

Answer: B

To answer this question, recall that a decibel is defined as ten times the log of the ratio of sound intensities:
(decibel measure) = 10 log (I / I_0) where I_0 is a reference intensity.

Therefore, in our case,
(decibels of first sound) = (decibels of second sound) + 10
10 log (I_1 / I_0) = 10 log (I_2 / I_0) + 10
10 log I_1 − 10 log I_0 = 10 log I_2 − 10 log I_0 + 10
10 log I_1 − 10 log I_2 = 10
log (I_1 / I_2) = 1
I_1 / I_2 = 10

This is consistent only with answer (B).
(Be careful not to get the two intensities confused with each other.)

64. The velocity of sound is greatest in:
(Average Rigor)

    A. Water

    B. Steel

    C. Alcohol

    D. Air

Answer: B

Sound is a longitudinal wave, which means that it shakes its medium in a way that propagates as sound traveling. The speed of sound depends on both elastic modulus and density, but for a comparison of the above choices, the answer is always that sound travels faster through a solid like steel, than through liquids or gases. Thus, the answer is (B).

65. A vibrating string's frequency is _____ proportional to the _____.
(Rigorous)

    A. Directly; Square root of the tension

    B. Inversely; Length of the string

    C. Inversely; Squared length of the string

    D. Inversely; Force of the plectrum

Answer: A

To answer this question, recall that
$f = (n\,v) / (2L)$ where f is frequency; v is velocity; L is length

and

$v = (F_{tension} / (m/L))^{1/2}$ where $F_{tension}$ is tension; m is mass; others as above

so

$f = (n / 2L) \, ((F_{tension} / (m/L))^{1/2})$

indicating that frequency is directly proportional to the square root of the tension force. This is consistent only with answer (A). Note that in the final frequency equation, there is an inverse relationship with the square root of the length (after canceling like terms). This is not one of the options, however.

66. Which of the following apparatus can be used to measure the wavelength of a sound produced by a tuning fork?
(Average Rigor)

    A. A glass cylinder, some water, and iron filings

    B. A glass cylinder, a meter stick, and some water

    C. A metronome and some ice water

    D. A comb and some tissue

Answer: B

To answer this question, recall that a sound will be amplified if it is reflected back to cause positive interference. This is the principle behind musical instruments that use vibrating columns of air to amplify sound (e.g. a pipe organ). Therefore, presumably a person could put varying amounts of water in the cylinder, and hold the vibrating tuning fork above the cylinder in each case. If the tuning fork sound is amplified when put at the top of the column, then the length of the air space would be an integral multiple of the sound's wavelength. This experiment is consistent with answer (B). Although the experiment would be tedious, none of the other options for materials suggest a better alternative.

67. The highest energy is associated with:
(Easy)

    A. UV radiation

    B. Yellow light

    C. Infrared radiation

    D. Gamma radiation

Answer: D

To answer this question, recall the electromagnetic spectrum. The highest energy (and therefore frequency) rays are those with the lowest wavelength, i.e. gamma rays. (In order of frequency from lowest to highest are: radio, microwave, infrared, red through violet visible light, ultraviolet, X-rays, gamma rays.) Thus, the only possible answer is (D). Note that even if you did not remember the spectrum, you could deduce that gamma radiation is considered dangerous and thus might have the highest energy.

68. **An object two meters tall is speeding toward a plane mirror at 10 m/s. What happens to the image as it nears the surface of the mirror?**
   (Rigorous)

   A. It becomes inverted.

   B. The Doppler Effect must be considered.

   C. It remains two meters tall.

   D. It changes from a real image to a virtual image.

Answer: C

Note that the mirror is a plane mirror, so the image is always a virtual image of the same size as the object. If the mirror were concave, then the image would be inverted until the object came within the focal distance of the mirror. The Doppler Effect is not relevant here. Thus, the only possible answer is (C).

69. **Automobile mirrors that have a sign, "objects are closer than they appear" say so because:**
   (Rigorous)

   A. The real image of an obstacle, through a converging lens, appears farther away than the object.

   B. The real or virtual image of an obstacle, through a converging mirror, appears farther away than the object.

   C. The real image of an obstacle, through a diverging lens, appears farther away than the object.

   D. The virtual image of an obstacle, through a diverging mirror, appears farther away than the object.

Answer: D

To answer this question, first eliminate answer choices (A) and (C), because we have a mirror, not a lens. Then draw ray diagrams for diverging (convex) and converging (concave) mirrors, and note that because the focal point of a diverging mirror is behind the surface, the image is smaller than the object. This creates the illusion that the object is farther away, and therefore (D) is the correct answer.

70. If an object is 20 cm from a convex lens whose focal length is 10 cm, the image is:
(Rigorous)

    A. Virtual and upright

    B. Real and inverted

    C. Larger than the object

    D. Smaller than the object

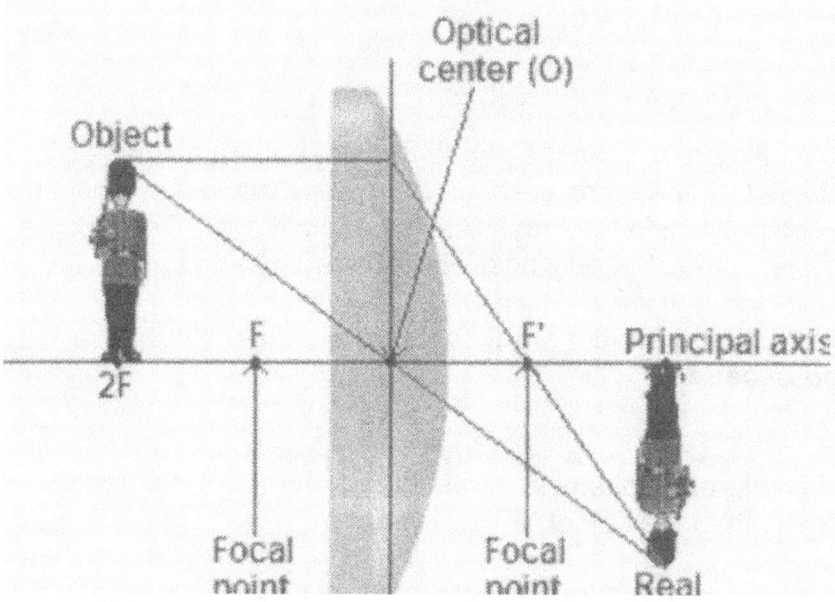

Answer: B

To solve this problem, draw a lens diagram with the lens, focal length, and image size.

The ray from the top of the object straight to the lens is focused through the far focus point; the ray from the top of the object through the near focus goes straight through the lens; the ray from the top of the object through the center of the lens continues. These three meet to form the "top" of the image, which is therefore real and inverted. This is consistent only with answer (B).

TEACHER CERTIFICATION STUDY GUIDE

71. **The constant of proportionality between the energy and the frequency of electromagnetic radiation is known as the:**
(Easy)

    A. Rydberg constant

    B. Energy constant

    C. Planck constant

    D. Einstein constant

Answer: C

Planck estimated his constant to determine the ratio between energy and frequency of radiation. The Rydberg constant is used to find the wavelengths of the visible lines on the hydrogen spectrum.

The other options are not relevant options, and may not actually have physical meaning. Therefore, the only possible answer is (C).

72. **Which phenomenon was first explained using the concept of quantization of energy, thus providing one of the key foundational principles for the later development of quantum theory?**
(Rigorous)

    A. The photoelectric effect

    B. Time dilation

    C. Blackbody radiation

    D. Magnetism

Answer: C

Although the photoelectric effect applied principles of quantization in explaining the behavior of electrons emitted from a metallic surface when the surface is illuminated with electromagnetic radiation, the explanation of the phenomenon of blackbody radiation, provided by Max Planck, was the first major success of the concept of quantized energy. Magnetism may be explained quantum mechanically, but such an explanation was not forthcoming until well after Planck's quantization hypothesis. Time dilation is primarily explained through relativity theory.

73. **Which statement best describes why population inversion is necessary for a laser to operate?**
    (Rigorous)

    A. Population inversion prevents too many electrons from being excited into higher energy levels, thus preventing damage to the gain medium.

    B. Population inversion maintains a sufficient number of electrons in a higher energy state so as to allow a significant amount of stimulated emission.

    C. Population inversion prevents the laser from producing coherent light.

    D. Population inversion is not necessary for the operation of most lasers.

Answer: B

Population inversion is a state in which there are a larger number of electrons in a particular higher-energy excited state than in a particular lower-energy state. When perturbed by a passing photon, these electrons may then emit a photon of the same energy (frequency) and phase. This is the process of stimulated emission, which, when population inversion is obtained, can produce something of a "chain reaction," thus giving lasers their characteristically monochromatic and highly coherent light.

74. **Bohr's theory of the atom was the first to quantize:**
    (Average Rigor)

    A. Work

    B. Angular Momentum

    C. Torque

    D. Duality

Answer: B

Bohr was the first to quantize the angular momentum of electrons, as he combined Rutherford's planet-style model with his knowledge of emerging quantum theory. Recall that he derived a "quantum condition" for the single electron, requiring electrons to exist at specific energy levels

75. **Two neutral isotopes of a chemical element have the same numbers of:**
    (Easy)

    A. Electrons and Neutrons

    B. Electrons and Protons

    C. Protons and Neutrons

    D. Electrons, Neutrons, and Protons

Answer: B

To answer this question, recall that isotopes vary in their number of neutrons. (This fact alone eliminates answers (A), (C), and (D).) If you did not recall that fact, note that we are given that the two samples are of the same element, constraining the number of protons to be the same in each case. Then, use the fact that the samples are neutral, so the number of electrons must exactly balance the number of protons in each case. The only correct answer is thus (B).

76. **When a radioactive material emits an alpha particle only, its atomic number will:**
    (Average Rigor)

    A. Decrease

    B. Increase

    C. Remain unchanged

    D. Change randomly

Answer: A

To answer this question, recall that in alpha decay, a nucleus emits the equivalent of a Helium atom. This includes two protons, so the original material changes its atomic number by a decrease of two.

77. Ten grams of a sample of a radioactive material (half-life = 12 days) were stored for 48 days and re-weighed. The new mass of material was:
(Rigorous)

   A. 1.25 g

   B. 2.5 g

   C. 0.83 g

   D. 0.625 g

Answer: D

To answer this question, note that 48 days is four half-lives for the material. Thus, the sample will degrade by half four times. At first, there are ten grams, then (after the first half-life) 5 g, then 2.5 g, then 1.25 g, and after the fourth half-life, there remains 0.625 g. You could also do the problem mathematically, by multiplying ten times $(½)^4$, i.e. ½ for each half-life elapsed.

78. Which of the following pairs of elements are not found to fuse in the centers of stars?
(Average Rigor)

   A. Oxygen and Helium

   B. Carbon and Hydrogen

   C. Beryllium and Helium

   D. Cobalt and Hydrogen

Answer: D

To answer this question, recall that fusion is possible only when the final product has more binding energy than the reactants. Because binding energy peaks near a mass number of around 56, corresponding to Iron, any heavier elements would be unlikely to fuse in a typical star. (In very massive stars, there may be enough energy to fuse heavier elements.) Of all the listed elements, only Cobalt is heavier than iron, so answer (D) is correct.

TEACHER CERTIFICATION STUDY GUIDE

79. **In a fission reactor, heavy water:**
(Average Rigor)

    A. Cools off neutrons to control temperature

    B. Moderates fission reactions

    C. Initiates the reaction chain

    D. Dissolves control rods

Answer: B

In a nuclear reactor, heavy water is made up of oxygen atoms with hydrogen atoms called 'deuterium,' which contain two neutrons each. This allows the water to slow down (moderate) the neutrons, without absorbing many of them. This is consistent only with answer (B).

80. Given the following values for the masses of a proton, a neutron and an alpha particle, what is the nuclear binding energy of an alpha particle?
    (Rigorous)

    Proton mass = 1.6726×10^{-27} kg
    Neutron mass = 1.6749×10^{-27} kg
    Alpha particle mass = 6.6465×10^{-27} kg

    A. 0 J

    B. 7.3417×10^{-27} J

    C. 4 J

    D. 4.3589×10^{-12} J

Answer: D

The nuclear binding energy is the amount of energy that is required to break the nucleus into its component nucleons. In this case, the binding energy of an alpha particle, which is composed of two protons and two neutrons, is calculated by first finding the difference between the sum of the masses of all the nucleons and the mass of the alpha particle. Using the equation $E = mc^2$ to find the energy in terms of the mass difference of 4.85×10^{-29} kg, and using the speed of light of about 2.9979×10^8 m/s, the result is the value given in answer D.

XAMonline, INC. 21 Orient Ave. Melrose, MA 02176

Toll Free number 800-509-4128

TO ORDER Fax 781-662-9268 OR www.XAMonline.com
GEORGIA ASSESSMENTS FOR THE CERTIFICATION OF EDUCATORS -GACE - 2007

PO#   Store/School:

Address 1:

Address 2 (Ship to other):
City, State Zip

Credit card number_____-_____-_____-_____   expiration_____
EMAIL _____
PHONE   **FAX**

13# ISBN 2007	TITLE	Qty	Retail	Total
978-1-58197-533-8	Basic Skills 200, 201, 202			
978-1-58197-528-4	Biology 026, 027			
978-1-58197-529-1	Science 024, 025			
978-1-58197-569-1	Physics 030, 031			
978-1-58197-531-4	Art Education Sample Test 109, 110			
978-1-58197-574-1	English 020, 021			
978-1-58197-545-1	History 034, 035			
978-1-58197-527-7	Health and Physical Education 115, 116			
978-1-58197-540-6	Chemistry 028, 029			
978-1-58197-534-5	Reading 117, 118			
978-1-58197-547-5	Media Specialist 101, 102			
978-1-58197-573-4	Middle Grades Language Arts 011			
978-1-58197-535-2	Middle Grades Reading 012			
978-1-58197-539-0	Middle Grades Science 014			
978-1-58197-543-7	Middle Grades Mathematics 013			
978-1-58197-546-8	Middle Grades Social Science 015			
978-1-58197-536-9	Mathematics 022, 023			
978-1-58197-549-9	Political Science 032, 033			
978-1-58197-544-4	Paraprofessional Assessment 177			
978-1-58197-542-0	Professional Pedagogy Assessment 171, 172			
978-1-58197-537-6	Early Childhood Education 001, 002			
978-1-58197-548-2	School Counseling 103, 104			
978-1-58197-541-3	Spanish 141, 142			
978-1-58197-538-3	Special Education General Curriculum 081, 082			
978-1-58197-530-7	French Sample Test 143, 144			
			SUBTOTAL	
	FOR PRODUCT PRICES GO TO WWW.XAMONLINE.COM		**Ship**	$8.25
			TOTAL	

www.ingramcontent.com/pod-product-compliance
Lightning Source LLC
Chambersburg PA
CBHW080538300426
44111CB00017B/2790